THE LOST ART OF WAR

THE
LOST ART
OF WAR

SUN TZU II

Translated with Commentary by
THOMAS CLEARY

HarperSanFrancisco
An Imprint of HarperCollins*Publishers*

HarperCollins Web Site: http://www.harpercollins.com

HarperCollins®, ▦ ®, and HarperSanFrancisco™ are trademarks of HarperCollins Publishers Inc.

FIRST HARPERCOLLINS PAPERBACK EDITION PUBLISHED IN 1997

Library of Congress Cataloging-in-Publication Data
Sun, Pin, 4th cent. B.C.
  [Sun Pin ping fa. English]
  The lost art of war / Sun Tzu II ; translated and presented by
Thomas Cleary.
  Includes bibliographical references.
  ISBN 0–06–251361–3 (cloth)
  ISBN 0–06–251405–9 (pbk.)
  1. Military art and science—Early works to 1800. 2. Strategy—
Early works to 1800. I. Cleary, Thomas. II. Title.
U101.S93513 1996   95–40157
355.02—dc20

97 98 99 00 01 ❖ HAD 10 9 8 7 6 5 4 3 2 1

CONTENTS

Introduction   I

The Lost Art of War: Sun Bin's *Art of War*   9

1 ▪ The Capture of Pang Juan   11
2 ▪ [Title Lost]   19
3 ▪ Questions of King Wei   25
4 ▪ Tian Ji Asks About Ramparts   39
5 ▪ Elite Troops   41
6 ▪ Timing Combat   47
7 ▪ Eight Battle Formations   51
8 ▪ Terrain and Security   53
9 ▪ Configurations of Force and Strategic Planning   59
10 ▪ Military Conditions   63
11 ▪ Practicing Selection   65
12 ▪ Killing Soldiers   69
13 ▪ Prolonging Energy   71
14 ▪ Official Posts   73
15 ▪ Strengthening the Military   85
16 ▪ Ten Battle Formations   87
17 ▪ Ten Questions   91
18 ▪ [Title Lost]   97
19 ▪ Distinctions Between Aggressors and Defenders   99
20 ▪ Experts   103
21 ▪ Five Descriptions and Five Courtesies   105
22 ▪ Mistakes in Warfare   107
23 ▪ Justice in Commanders   III
24 ▪ Effectiveness in Commanders   113

25 ▪ Failings in Commanders   115
26 ▪ Losses of Commanders   117
27 ▪ Strong and Weak Cities   121
28 ▪ [Title Lost]   123
29 ▪ [Title Lost]   125
30 ▪ Surprise and Straightforwardness   127

Leadership, Organization, and Strategy:
How Sun Tzu and Sun Tzu II
Complement Each Other   131

INTRODUCTION

FEW THINGS might seem as unlikely as ancient Chinese warriors rising up in modern American business schools, corporate boardrooms, and Hollywood movies, but there they are.

The rise of the new China, the power of the global marketplace, the intensification of international competition and rivalry—all of these elements of contemporary affairs may contribute in some way to present-day interest in ancient classics of strategy and conflict management.

This literature, however, may have yet another function—perhaps not as apparent, but no less important than what emerges from its use in business, diplomacy, and warfare. While the study and application of strategic thinking in these areas may be necessary to secure the economic, political, and territorial underpinnings of democracy, these cannot guarantee freedom unless individuals and peoples have the right and the opportunity to recognize and understand all of the operative influences in their lives and on their minds—not only those that happen to be, or are made to appear, most evident to everyday awareness. Only thus is true freedom of choice possible in a real, practical sense.

One of the most important functions of strategic literature in the public domain today, therefore, may be to enhance the general understanding of power and its uses and abuses. By understanding power—how configurations of power work, how masses of people are influenced, how individuals and peoples become vulnerable to internal contradictions and external aggression—it is possible to objectively and truthfully assess the operation of the world we live in—

I

and, we might hope, to learn to avoid the abuses of power to which the massive, impersonal infrastructures of modern life are inherently prone.

The most famous of ancient Chinese strategic manuals is *Sun Tzu's Art of War* by Sun Wu, an outstanding military consultant of China's Warring States era. Somewhat more than a hundred years later, a lineal descendant of this Sun Tzu, or "Master Sun," also rose to prominence as a distinguished strategist. This was Sun Bin, whose name means Sun the Mutilated. He became Sun Tzu II, the second Master Sun, whose own *Art of War* was only known in part until a nearly complete, if somewhat damaged version recorded on bamboo strips was recently discovered in an ancient tomb.

It is not accidental that the great classics of Chinese political and military strategy emerged from the chaos and horror of the Era of the Warring States. The conditions of the time, and their effect on the morale and philosophy of the people, are graphically described in the traditional anthology known as *Strategies of the Warring States:*

> Usurpers set themselves up as lords and kings; states that were run by pretenders and plotters established armies to make themselves into major powers. They imitated each other at this more and more, and those who came after them also followed their example. Eventually they overwhelmed and destroyed one another, conspiring with larger domains to annex smaller domains, spending years at violent military operations, filling the fields with blood.
>
> Fathers and sons were alienated, brothers were at odds, husbands and wives were estranged. No one could safeguard his or her life. Integrity disappeared. Eventually this reached

the extreme where seven large states and five smaller states contested each other for power. This all happened because the warring states were shamelessly greedy, struggling insatiably to get ahead.[1]

According to ancient documents, Sun Bin studied warfare along with a certain Pang Juan, who later became a high-ranking military leader. Their teacher was supposed to have been the mysterious sage Wang Li, known as the Master of Demon Valley, one of the most redoubtable strategic theorists of all time.

Reputedly the author of *The Master of Demon Valley*, the most sophisticated of all strategic classics, Wang Li was a Taoist recluse. According to Taoist records, while certain students of his became prominent strategists active in the melees of the Warring States, the Master of Demon Valley tried in vain to induce them to use their knowledge to convert the warlords to Taoism rather than to hegemonism.

Legend has it that the Master of Demon Valley lived for hundreds of years. While this sort of fable is ordinarily connected to esoteric Taoist life-prolonging theories and practices, in this case it may allude to the maintenance of a highly secretive hidden tradition. Such a pattern of esoterism in the preservation and transmission of potentially dangerous knowledge was typical, to be sure, of ancient Chinese schools, particularly under conditions of social disorder.

The book known as *The Master of Demon Valley* most certainly contains much that is not in the lore of the first Sun Tzu's popular *Art of War*.[2] The book of Sun Tzu II, Sun the Mutilated, also bears the stamp of extraordinary knowledge that may in fact derive from the school of the Master of

Demon Valley. The reason for secrecy, which also explains the extremely cryptic language used in recording such texts, is made clear in the conclusion to *The Master of Demon Valley:* "Petty people imitating others will use this in a perverse and sinister way, even to the point where they can destroy families and usurp countries."[3] This is the tradition of Sun the Mutilated.

After completing his study of tactical strategy with the mysterious Master of Demon Valley, Sun Bin's schoolmate Pang Juan was hired by the court of the state of Wei, where he was appointed to the rank of general. Concerned that his own abilities were unequal to those of Sun Bin, General Pang Juan devised a plan to remove him from the scene.

To encompass his rival's downfall, Pang Juan had Sun Bin invited to Wei as if to consult with him. When Sun Bin arrived, however, Pang Juan had him arrested as a criminal. Falling victim to the plot and condemned as a convict, Sun Bin had both feet amputated and his face tattooed. This is why he came to be known as Sun the Mutilated. Such punishments were designed to reduce people to the status of permanent outcasts.

Sun Bin, however, was evidently undaunted by this setback. Perhaps he considered himself most at fault for having fallen into Pang Juan's trap. In any case, while he was a convict/slave, Sun Bin gained a private audience with an emissary of the state of Qi who was passing through Wei en route to the state of Liang. Taking advantage of his opportunity, Sun Bin astounded the ambassador with his extraordinary knowledge of strategy and warfare.

Recognizing the value of such a mind, the emissary smuggled Master Sun out of Wei into his own state of Qi. Now the ruler of Qi wanted to make Sun Bin a general, but he pleaded infirmity on account of having been judicially maimed. As a result, the strategist was instead appointed military consultant to the great general Tian Ji.

Sun Bin's skill in the classical strategy of his ancestor Sun Wu and his teacher Wang Li is illustrated by a famous story of his service to Qi, immortalized in the popular *Extraordinary Strategies of a Hundred Battles* by the great Ming dynasty warrior-scholar Liu Ji:[4] When two states attacked a third, the victimized state appealed to the state of Qi for help. The general of Qi asked Sun Bin for advice. Master Sun said, "The aggressor armies are fierce and think little of your army, which they regard as cowardly. A good warrior would take advantage of this tendency and 'lead them on with prospects of gain.'

"According to *The Art of War*, struggling for an advantage fifty miles away will thwart the forward leadership, and only half of those who chase prospects of gain twenty-five miles away will actually get there. Have your army enter enemy territory and make thousands of campfires; on the next day have them make half that number of campfires, and on the day following that have them make half again that number."

The general had his army do as Sun Bin advised. The enemy general was deceived by this maneuver, delighted to hear that the number of campfires was dwindling day by day, assuming that the men of Qi were defecting. He said, "I knew the soldiers of Qi were cowards—they've only been in our territory for three days now, and more than half their army has run away!"

As a result of this misperception, the enemy general left his own infantry behind and rushed in pursuit of the Qi army with nothing but a personal force of crack troops. Calculating

the speed of their pursuit, Master Sun the Mutilated con-
cluded that by nightfall his opponents would reach an area of
narrow roads difficult to pass through, a place suitable for
ambush.

Now Master Sun had a large tree felled and stripped of
bark. Then he wrote on the bare log, "The general of Wei
will die at this tree," and had it placed on the road where the
aggressor troops of Wei would pass that night. Then Master
Sun had several thousand expert archers conceal themselves
near the road.

When the general of Wei, Sun Bin's opponent, came to
the place where the stripped log lay across the road, the general had a torch kindled for light to read the writing on the
log. Before he had finished reading the inscription stating
that he himself would die right there that night, the archers
placed by Sun Bin loosed their arrows at the sight of the
torch, throwing the enemy into a panic. Realizing he had
been outwitted and his troops beaten, the general of Wei
committed suicide.

Thus the tactics of Master Sun the Mutilated encom-
passed victory at minimal cost. This is one of the cardinal
principles of the science.

· · ·

Sun Bin's book of strategy, long known by fragments and
only recently discovered in the same ancient tomb in which
the previously unknown version of Sun Tzu's *Art of War* was
found, was most likely compiled by disciples. Like other
works of pre-imperial China, this text appears to be a collec-
tion of aphorisms and analects, largely organized in a lateral
associative manner rather than in a linear progressive order.

The book begins with a cryptic story of Sun Bin's ultimate triumph over his old nemesis, Pang Juan, using the strategy of misdirection advocated by his ancestor Sun Wu. This account, which is intrinsically obscure and subject to different readings and interpretations, mainly serves to establish the superior mastery of Master Sun the Mutilated. Undoubtedly, it was for this purpose that the followers of Sun Bin introduced their account of their master's teachings with this story.

The text goes on to record some of master Sun Bin's conversations with the king and the general Tian Ji of Qi, to whom he acted as consultant. After that, the book proceeds to deal with a series of related topics, focusing on hard-core issues of tactics and strategy. The book of Sun Bin is a composite text, based on a Chinese sense of holistic order rather than a Hellenic sense of logic. Thus it covers a wide range of subjects as a totality, without the linear progression familiar to scholastic Western thought. As an esoteric work of military strategy, furthermore, it is by nature highly secretive, recorded in extremely difficult language, often substituting homonymic characters in a kind of cryptic code.

A century deeper into the chaos of Warring States China, while retaining the ancient moral foundations of Sun Wu's *Art of War,* Sun Bin went that much further than his distinguished ancestor in detailing practical tactics. Like what can be found in other strategic manuals, moreover, Sun Bin's methods are represented by structures that operate as metaphors for events and activities other than warfare, in the domains of government, diplomacy, business, and social action.

There is a Chinese saying, "The wise see wisdom, the good see goodness." How people interact with powerful and secretive lore like *The Art of War* is held to reveal something about their inner character—and this function itself is a classic maneuver of strategic art.

NOTES

1. Cited in Thomas Cleary, *The Art of War* (Boston: Shambhala, 1988). It has often been noted how much the highly competitive contemporary international marketplace resembles the Era of the Warring States.

2. For a complete translation of the esoteric classic of strategy known as "The Master of Demon Valley," see Thomas Cleary, *Thunder in the Sky: On the Acquisition and Exercise of Power* (Boston: Shambhala, 1993). "The Master of Demon Valley" contains much more of the psychological, social, and political dimension of strategy in action than does Sun Tzu's renowned *The Art of War,* which is more strictly military. "The Master of Demon Valley" illustrates the Taoist ideas and techniques in traditional strategy more prominently than do either of the Suns, Sun Tzu I (Sun Wu) or Sun Tzu II (Sun Bin). Taoist legend about Wang Li, the Master of Demon Valley, has the maestro weeping over his wayward disciples who became famous military strategists but failed to convert the lords to the moral dimensions of the Tao.

3. Cleary, *Thunder in the Sky*, 76.

4. For Liu Ji's own commentary on how this story illustrates the art of war, see Thomas Cleary, *Mastering the Art of War* (Boston: Shambhala, 1989), 96–98.

THE
LOST ART
OF WAR:

▪ Sun Bin's *Art of War* ▪

In the past, when the ruler of Liang was going to attack Handan, the capital of Zhao, he sent his general, Pang Juan, to Chuqiu with 80,000 armed troops.

▪ ▪* Sun Bin's *Art of War* begins, appropriately enough, with the story of how he vanquished his arch rival, Pang Juan. The shifting alliances of China's Era of the Warring States created enormous confusion and uncertainty; and the chaotic and unpredictable nature of the political scene was ruthlessly exploited by civil and military leaders as well as by freelance consultants. To read ancient Chinese war stories like this, which do not necessarily follow a straight line, and whose action is inherently and perhaps deliberately confusing, observe the relationships and interactions as well as the overall "plot."

When King Wei of Qi (B.C.E. 356–320) heard of this, he had his general, Tian Ji, lead 80,000 armed troops to the rescue.

Pang Juan attacked the capital of Wey. General Tian Ji asked Master Sun Bin, "What should I do if I do not rescue Wey?"

▪ ▪ When Sun Bin was offered a generalship by King Wei, he refused the honor; ostensibly he did so on account of his physical disabilities, but perhaps for other strategic reasons as well.

---

*This symbol (▪ ▪) denotes commentary by the author.

Instead, he was appointed military adviser to the top general, Tian Ji.

Master Sun said, "Please attack Rangling to the south. The walled city of Rangling is small, but its province is large, populous, and well armed. It is the essential military zone in eastern Wei; it is hard to besiege, so I would make a show of confusion. Were I to attack Rangling, I would have the state of Song to my south and the state of Wey to my north, with Shiqiu right in the way, so my supply lines would be cut off; I would thus make a show of incompetence."

■ ■ Two essential principles of conflict outlined by Sun Bin's distinguished ancestor Sun Wu are illustrated here. One concerns the way to draw an enemy out of a secure position by attacking a place of strategic value, somewhere that the opponent is sure to go to the defense. The Master of Demon Valley, another Warring States–era strategist, also described the practice of drawing people out in interactions by observing their structures of psychological defense. The exercise of tact, as well as apparent lack of tact, may both be used in strategic encounters as means of observing reaction patterns in actual or potential allies or adversaries, either from a "closed" undercover position (disguised as tact) or from the shifting, roving, deceptively "open" position of the provocateur (disguised as lack of tact).

The second principle illustrated here is the use of deceit for strategic advantage in hostile situations. In this case, the particular avenue of deception employed is that of deliberately giving the appearance of confusion and lack of skill in order to make an enemy contemptuous, complacent, and therefore careless, rendering the enemy vulnerable to counterattack. Thus, in situations where strategy is the paramount guide of affairs, it is customary not to take anything naively at face value; yet it is therefore crucial to attain the intelligence and balance to avoid becoming excessively imaginative and lapsing into self-defeating paranoia.

So General Tian broke camp and raced to Rangling.

Subsequently, General Tian summoned Master Sun and asked, "Now what should I do?"

Master Sun said, "Which of the grandees of the cities are ignorant of military affairs?"

General Tian replied, "Those in Qi-cheng and Gao-tang."

▪▪ Here, the least competent are being selected because they are to be pawns in a larger game. Their role, which they are to play unawares, is to keep up the appearance of lack of skill and intelligence on the part of the leadership. They are also being set up to be sacrificed for a larger cause; this is one of the notorious Thirty-Six Strategies.

Master Sun said, "Get these grandees to take charge of defense of their cities. Fan out and then close in on all sides, circling around. Fanning out and then closing in with a battlefront, circle around to station camouflaged soldiers. With your vanguard forceful, have your main force continuously circle around and strike the opponent from behind. The two grandees can be sacrificed."

So Qi-cheng and Gao-tang were separated into two units, and a direct assault on Rangling was made in an attempt to swarm the city. Camouflaged ambushers circled around and struck from behind; Qi-cheng and Gao-tang were routed, succumbing to the strategy.

■ ■ Master Sun advised General Tian to draw the opponent in by this show of incompetence, then come upon the enemy from behind with a sneak counterattack. In this case, the opposing side added another layer of the same strategy, thereby succeeding in gaining an immediate aim, yet also falling into the trap of Master Sun Bin.

The master strategist had already taken this possibility, or indeed likelihood, fully into account; the purpose of the siege ordered by Sun Bin was not to win its ostensible aim, but to render his opponent vulnerable. When you see a target, and you see people aiming for the target, try to see what the actual purpose—in effect, that is—might be: the target, the aiming, the attention of onlookers, whatever might possibly be gained from a combination of some or all of these under prevailing conditions, or possibly something else again.

General Tian summoned Master Sun and said, "I failed in the siege of Rangling, and also lost Qi-cheng and Gao-tang, which were defeated strategically. Now what should I do?"

Master Sun said, "Please send light chariots galloping west to the city of Liang, to enrage them; follow up with your troops split up to make it look as if they are few in number."

■ ■ It may seem dangerous to enrage an enemy; and indeed it may be, so it is imperative to examine the specific conditions of the moment. The purpose of strategic use of anger in this way is to blunt the effectiveness of an opponent by making him lose his head and expend his energy wildly, without deadly concentration. The great heavyweight champion boxers Jack Johnson and Muhammad Ali were particularly renowned for their consummate skill in this art. In some cultures, what Americans call "in your face" manners or mannerisms are in fact either active examples or not-quite-dead relics of this type of strategic behavior, developed within the context of social conditions that tend to force people to assume such behavior, if for no other reason than instinctive self-preservation.

There is another function of anger noted in classics of Chinese strategy, that of enraging one's own forces against a powerful enemy in order to add to the psychological momentum of one's attack and surety of one's defense. Ancient Norse berserkers, from whom we derive the word berserk, practiced this on themselves.

In American pugilism, the legendary middleweight boxing champion Stanley Ketchel was famous for this berserker-like practice. Powerful enough in his rage to floor the great heavyweight champion Jack Johnson himself, Ketchel seems to have been one of the few truly violent young men who ostensibly succeeded in this profession. His habitual indulgence in his own passions, however, and his consequent forgetfulness of passion as a double-edged sword—capable of making men killers as well as buffoons—was precisely what led to his own early death. One of the most murderous of men ever to box under the Marquis of Queensbury rules,

Stanley Ketchel was only twenty-four years old when he was shot to death by an angry cuckold.

Nowadays, the practice of enraging one's troops against an enemy seems to be widely used. Film clips of such training procedures are even to be seen on national educational television from time to time. The use of such tactics on soldiers of a highly diverse nation with ancestral roots all over the world has very serious drawbacks. Already evident on the battlefield, these drawbacks are particularly glaring when viewed in the total context, including the relationship between civil society and its own military. It may be for this reason that strategic lore in China, a land of great regional diversity, generally speaks more of enraging others than enraging one's own minions or allies against others.

In this connection, it may be noteworthy that Sun Bin's book on strategy does not vilify his arch enemy Pang Juan, the man responsible, out of paranoid jealousy, for having Sun Bin's feet amputated and his face tattooed as a criminal. The book contains no hint, for example, of Sun Bin attempting to enrage his advisee General Tian against Pang Juan; nor does he leave a string of uncomplimentary epithets in his work to curse his malefactor until the end of time. This does not mean Sun Bin had no feelings, of course; only that he kept his professional cool in spite of what emotions he might have felt at heart. In such cases, it should be noted, this virtue is not necessarily a moral virtue; it may not be any more than a strategic virtue. To make this distinction clearly is itself a strategic advantage, for it diminishes vulnerability to confusion, which can occur and can also be exploited in any domain, dimension, or form.

General Tian did as Master Sun recommended, and General Pang did in fact come by forced march, leaving equipment behind. Master Sun attacked him relentlessly at Guiling, and captured General Pang.

■ ■ All along, Sun Bin's strategy had been aimed at getting General Pang to overreach and expose himself carelessly, in the process also tiring his troops out and tricking them into coming to the fray too lightly equipped.

That is why it was said that Master Sun was a consummate expert at his business.

■ ■ In sum, Sun Bin got General Tian to employ classic tactics of misdirection, inward and outward deception, giving up something in order to get something more valued, strategically exploiting the quirks, weaknesses, and shortcomings of human neurological, perceptive, and emotional functions to achieve a specific purpose. The Taoist classic *Tao Te Ching* says, "The Tao is universal; it can be used for the right or the left." This means that natural laws can be caused to operate in ways that may be amoral or even immoral as well as in ways that may be moral. It is for this reason that development of both character and perceptivity was in ancient times a traditional Taoist requisite for learning to practice the Tao in one's own life.

When Master Sun met King Wei, he said, "A militia is not to rely on a fixed formation; this is the way transmitted by kings of yore."

▪ ▪ The reasons for not relying on a fixed formation are both defensive and offensive. In terms of defense, predictability means vulnerability, since an adversary who knows how you will react to a given situation will be able to take advantage of this knowledge to scheme against you. If you are not so predictable, on the other hand, not only will adversaries be unable to pinpoint their targets, but their attention will be weakened by dispersal.

When opponents cannot predict what you will do, they cannot act against you with inevitable effect; and when they realize they cannot predict what you will do, they have to be more watchful without knowing quite what they are watching for, thus exhausting the energy of attention and progressively diluting its effectiveness.

This leads naturally into the offensive aspect of unpredictability; enemies who cannot tell when or where you might act are thereby prevented from preparing a sure defense. Their attention is thus necessarily spread more thinly, and their mental energy naturally wanes on account of the added burden. The buildup of constantly mounting anxiety accelerates this process, and aggressively unpredictable behavior that is fundamentally intended to increase tension succeeds doubly in its function by the added tension inherent in futile reactions to misdirection.

"Victory in war is a means of preserving perishing nations and perpetuating dying societies; failing to win in war is how territory is lost and sovereignty threatened. This is why military matters must be examined."

■ ■ In her small but powerful book *Prisons We Choose to Live Inside* (1987), Doris Lessing observes the tragedy of those who believe in freedom and peace but inhibit their own liberation and fulfillment by refusing to examine the mechanisms of oppression and war. If we wish to remedy malignant conditions, she argues, we need to understand those conditions and how they affect us. This is precisely the logic of the ancient *Art of War:* Know your enemy, know yourself, know where you are, know what is going on.

The idea of smashing the mechanisms of oppression and war may be emotionally stimulating, but it is both childish and exactly contrary to strategic common sense because it falls into the simplest of traps. Dismantling these mechanisms, not dreaming of smashing them, is a sounder and more intelligent approach—provided it is not just a calmer dream and is actually empowered with knowledge and understanding of their designs and operations.

These reasons for studying the art of war are already set forth with simple clarity by classical philosophers of China, whom Sun Bin is informally citing here. In sum, the logical purpose of learning about the workings of conflict is to be able to preserve innocents from aggression, oppression, and destruction. This is considered to be an intelligent and civilized extension of the natural instinct of self-preservation.

This applies not only to warfare in a literal sense, but to all fields of competition and contention, all domains of hostility and conflict. Before we can fairly understand what we might be able to do about anything, we need to see what aims are being served and what means are being employed. Without this mental equipment, we are likely to become unable to react to trying situations in any but emotionally overcharged but pragmatically inefficient ways.

"Those who enjoy militarism, however, will perish; and those who are ambitious for victory will be disgraced. War is not something to enjoy, victory is not to be an object of ambition."

■ ■ The Taoist classic *Tao Te Ching* says, "Fine weapons are implements of ill omen: People may despise them, so those who are imbued with the Way do not dwell with them." The pacifism expressed here is not sentimental or naive; note that the text says people "may despise" weapons, not that people "do" despise weapons, or that people "all" despise weapons. Weaponry fetishes are well documented throughout the world from ancient to modern times, and contemporary sociological and psychological researches have testified to forms of this phenomenon so comparatively subtle as to be normally unidentified as such. Not being adequately described or identified as such in everyday consciousness, these influences therefore pose a more insidious threat to human stability than grosser and more readily identifiable forms of weaponry fetishism.

The *Tao Te Ching* continues, "Weapons, being instruments of ill omen, are not tools of the cultured, who use them only when unavoidable. They consider it best to be aloof; they win without beautifying it. Those who beautify it enjoy killing people." Also, "The good are effective, that is all; they do not

presume to grab power thereby. They are effective but not
conceited, effective but not proud, effective but not arrogant.
They are effective when they have to be, effective but not co-
ercive." These passages of the quintessential Taoist classic re-
flect with ample clarity the pristine Taoist inspiration of Sun
Bin's concept of the proper place of warfare in human affairs
as illustrated in the introduction to this chapter of his classic
manual, on strategic advice to a king.

**"Act only when prepared. When a citadel is small and
yet its defense is firm, that means it has supplies. When
there are few soldiers and yet the army is strong, that
means they have a sense of meaning. If they defend without supplies or fight without meaning, no one in the
world can be firm and strong."**

■ ■ To act only when prepared is the cardinal rule of all martial
arts and strategic action, and indeed of all business and cre-
ative endeavor. This perennial admonition is followed by a
series of diagnostic guidelines, because proper preparation is
only possible with knowledge of conditions for or against
which preparation is being made.

In this particular case, insofar as his remarks are introduc-
tory, Master Sun takes a general approach and summarizes the
main parameters according to which conditions can be usefully described: the material and the mental or moral. As is
often the case, this is simply a matter of common sense, not
only in warfare, but in any constructive activity. It is necessary
to have a concentrated sense of purpose to make effective use
of material resources, and it is necessary to have adequate
wherewithal to sustain the effort to actualize the aim.

"When Yao ruled the land in antiquity, there were seven instances where royal decrees were rejected and not carried out; two among the eastern tribes, four in the heartland of China, . . . Nothing could be gained from just letting things go, so Yao fought and won, and established himself strongly, so that everyone submitted.

"In high antiquity, Shennong warred against the Fusui tribe, the Yellow Emperor warred on the region of Shulu. Yao struck down the Gong Gong people, Shun struck down . . . and drove off the San Miao tribes. Tang banished a despot, King Wu struck down a tyrant. The Yan tribe of the old Shang confederacy rebelled, so the Duke of Zhou overcame it.

"Therefore it is said that if your virtue is not comparable to the Five Emperors of Antiquity, your ability is not comparable to the Three Kings of old, and your wisdom is not comparable to the Duke of Zhou, even if you say you are going to build up humaneness and justice and use ritual and music to govern peacefully, thus putting a stop to conflict and depredation, it is not that Yao and Shun did not want to be thus, but that they were not able to do so; that is why they marshaled warriors to rectify matters."

■ ■ Here Master Sun is also following classical tradition in citing events from the legends of ancient cultural heroes to underscore the need to understand the art of war even in just and peaceful societies. Shennong was a prehistorical leader, associated with the development of agriculture, horticulture, and herbal medicine. The Yellow Emperor, who is supposed to have reigned in the twenty-seventh century B.C.E., is one of the most important figures of Taoism, believed to have studied and collected a broad spectrum of esoteric knowledge. Yao and Shun, ancient kings of the twenty-fourth and twenty-third centuries B.C.E., are depicted as paragons of just rulership. King Tang was the founder of the Shang/Yin dynasty in the eighteenth century B.C.E.; King Wu and the Duke of Zhou were founders of the Zhou (Chou) dynasty in the late twelfth century B.C.E. Yao, Shun, Tang, Wu, and the Duke of Zhou were particularly revered in the Confucian tradition, which emphasized justice and humaneness in government and public service.

The Five Emperors and Three Kings, although differently named and listed in various traditional sources, collectively refer to legendary leaders symbolizing prototypes of wisdom and humanity in Taoist, Confucian, and other Chinese political traditions. The reasoning Sun Bin is using here in citing these images, echoed in eminent Taoist classics such as *The Masters of Huainan* and *Wen-tzu,* is this: Since even great leaders of the past renowned for the benevolence of their regimes were not able to avoid hostilities, it follows that rulers of later times, no matter how good their intentions, cannot afford to ignore the science of conflict management and ignore the arts of strategy. *The Masters of Huainan,* a comprehensive Taoist classic of the second century B.C.E., illustrates this humanistic approach to the issue of military preparedness: "Those who used arms in ancient times did not do so to expand their territory or obtain wealth; they did so for the survival and continuity of nations on the brink of destruction and extinction, to settle disorder in the world, and to get rid of what harmed the common people."

King Wei of Qi asked Master Sun about military operations in these terms: "When two armies are a match for each other and their commanders are at a standoff, with both sides holding firm and neither willing to make the first move, what should be done about this?"

Master Sun replied, "Test the other side by means of light troops, led by a brave man from the lower echelons. Aim to cause a setback, not to gain a victory. Create a hidden front to harass their flanks. This is considered great success."

■■ The purpose of aiming to cause a setback rather than gain a victory is to test the strength of the opponent without revealing the depth of one's own resources. The hidden front stands behind the dummy test front, awaiting the adversary's reaction.

This sort of maneuver takes places in all sorts of interactions. In Japanese, the overt content of an interpersonal transaction is called the *omote,* which means "front," or *tatemae,* which means "setup"; while the ulterior motive is called the *ura,* meaning "back," or *honne,* meaning something like "true voice." While the strategic use of gaps between overt expression and covert intention is probably universal negotiating practice, cultural differences in the manner of its operation and perception may obscure relevant parallels in apparently different behaviors.

Description of this tactic is useful for defensive purposes, because alertness and perceptivity can be enhanced simply by keeping it in mind. When in adversarial, disadvantageous, or simply unfamiliar circumstances, it is useful to remember that one and the same conversation or confrontation can simultaneously accomplish two or more purposes. What it all really means in actual effect depends on the way in which the total transaction is understood and the manner in which each party perceives and reacts to the other's overt moves and covert intentions. Sun Bin's distinguished predecessor, Sun Wu, wrote, "If you know yourself and also know others, you will not be endangered in a hundred battles."

King Wei asked, "Are there proper ways to employ large and small forces?"

Master Sun replied, "There are."

King Wei asked, "If I am stronger and more numerous than my enemy, what should I do?"

Master Sun answered, "This is the question of an intelligent king. When your forces are larger and more powerful, and yet you still ask about how to employ them, this is the way to guarantee your nation's security. Give the command for an auxiliary force. Disarray the troops in confused ranks, so as to make the other side complacent, and they will surely do battle."

■ ■ Complacency undermines strength, so the powerful can retain their power by avoiding complacency, while encompassing the downfall of adversaries by projecting such an image of incompetence as to induce complacency and contempt. Sun Wu wrote, "Even when you are solid, still be on the defensive; even when you are strong, be evasive."

King Wei asked, "When the enemy is more numerous and stronger than I, what should I do?"

Master Sun said, "Give the command for a retractable vanguard, making sure to hide the rear guard so the vanguard is able to get back safely. Deploy the long weapons on the front lines, the short weapons behind, with mobile archers to help the hard-pressed. Have the main force remain immobile, waiting to see what the enemy can do."

■ ■ The function of the vanguard is to harass the adversary, whether to induce exasperation, to draw a counterattack that would leave the opponent vulnerable, or to induce the enemy to divide and split off or otherwise abandon a position or configuration of power. The rear guard, naturally, is there to cover and back the vanguard up, while the main force lies in wait to follow up on any confusion or weakening of the opponent's power.

This passage also provides a useful metaphor for resource allocation in challenging situations, or when dealing with intractable problems. The vanguard is research, the rear guard is development, the main force is the existing infrastructure and resource allocation already in place.

The vanguard has to be "retractable" in that effective research needs to be flexible and adaptive, ready to start anew in fresh directions as conditions require. Fixed commitments that are unresponsive to changing conditions, like a vanguard that cannot be withdrawn, are more vulnerable to being compromised by the vagaries of the unpredictable and the unforeseen.

In a highly competitive environment, development as the "rear guard" backing up vanguard research is kept "hidden," or secret, so that everyone is not doing the same thing at the same time. This makes constructive, evolutionary competition possible, while helping to maintain a relatively open space in which public opinion can be expressed.

The "main force" of existing infrastructure (including abstract, conceptual infrastructures of culture) remains immobile in the sense that it requires a degree of stability in order to function effectively, yet must "see what the enemy will do" in order to evolve the organs and operations needed by the society, the company, and the individual to manage challenging or threatening situations.

King Wei asked, "Suppose the enemy comes out when I go out, and I do not yet know whose numbers are greater; what should I do?"

■ ■ The reply to this question is missing. Based on relevant materials in this and related texts, it might be surmised that Master Sun would be likely to have recommended tactics designed to feel out the opponent while concealing one's own strengths.

King Wei asked, "How should one attack desperadoes?" Master Sun answered, ". . . [Wait] until they find a way to live."

■ ■ The most ancient Chinese classic, the *I Ching,* or *Book of Changes,* is the first to outline this particular strategy of not driving opponents to deadly desperation. There, this is symbolized by a king on a hunt using only three chasers, leaving one corner of the dragnet open in order to give the prey a fighting chance to escape.

The principle is that a "cornered rat" may turn on its pursuer with inconceivably deadly force if it is driven to a frenzy in absolute despair. Given a chance to escape, the reasoning goes, vanquished opponents will not become embittered diehards and need not be imprisoned, suppressed, or exterminated to achieve and maintain peace and social order in the aftermath of conflict.

King Wei asked, "How do you attack equals?"

Master Sun replied, "Confusing them and splitting them up, I concentrate my troops to pick them off without the enemy realizing what is going on. If the enemy does not split up, however, settle down and do not move; do not strike where there is doubt."

▪ ▪ Confusing and splitting an opponent's force is done in order to diffuse and blunt the energy of an attack as well as to compromise the security of the enemy's defense. Striking the diffuse with concentrated force is a way to shift the balance of strategic factors in one's own favor when facing an adversary of equal size and strength. Concentration and diffusion are both mental and physical, applying to attention, momentum, force, numbers, and material resources. The existence of doubt means that there is diffusion of attention, resulting in loss of concentration; therefore it is recommended that no action be initiated in doubtful situations.

King Wei asked, "Is there a way to strike a force ten times my size?"

Master Sun replied, "Yes. Attack where they are unprepared, act when they least expect it."

■ ■ This tactic is taken directly from *The Art of War* by Sun Wu (Sun Tzu). The idea is that greater power and resources do not guarantee tactical superiority if they are not effectively employed. The purpose of deception as a strategic art, therefore, is to prevent adversaries from using their aggressive and defensive capacities accurately.

King Wei asked, "When the terrain is even and the troops are orderly, and yet they are beaten back in an engagement, what is the reason?"

Master Sun said, "The battlefront lacked an elite vanguard."

■ ■ The function of an elite vanguard is to harass, split up, confuse, and otherwise soften up the adversary.

King Wei asked, "How can I get my people to follow orders as an ordinary matter of course?"

Master Sun said, "Be trustworthy as an ordinary matter of course."

■ ■ If ever there was a golden key to the art of leadership, perhaps this is it: To get people to follow orders as a matter of course, be trustworthy as a matter of course. The practical philosopher Confucius is on record as observing that people will not obey leaders they do not trust, even if they are coerced; whereas they will follow leaders they do trust, even when nothing is said.

King Wei exclaimed, "Excellent words! The configurations of warfare are inexhaustible!"

■ ■ Inexhaustibility of configurations means endless adaptation. When surprise tactics are repeated over and over, they become conventionalized and lose their strategic value. When conventional tactics are altered unexpectedly according to the situation, they take on the element of surprise and increase in strategic value. Thus it is said that the surprise becomes conventional, while the conventional becomes a surprise.

General Tian Ji asked Master Sun, "What causes a militia trouble? What thwarts an opponent? What makes walls impregnable? What makes one miss opportunities? What makes one lose the advantage of the terrain? What causes disaffection of people? May I ask if there are underlying principles governing these things?"

Master Sun replied, "There are. Terrain is what causes militias trouble, narrow passages are what thwart opponents. Thus it is said, 'A mile of swamp is a commander's nightmare; . . . to cross over, they leave their full armor behind.' That's why I say that it is the terrain that troubles an army, narrow passages that thwart an opponent, barbed wire that makes walls impenetrable, ..."

■ ■ In the original, parts of Sun Bin's reply are missing. The master's explanations of how opportunities are missed and how disaffection occurs are lost, but similar themes appear elsewhere in classical strategic lore, being critical issues of leadership.

Generally speaking, opportunities are lost through misinformation or lack of information, faulty evaluation of intelligence, lack of courage or initiative, indolence, preoccupation, or similar flaws in basic management. The great civil and military leader Zhuge Liang said, "There are three avenues of

SUN TZU II

opportunity: events, trends, and conditions. When opportunities occur through events but you are unable to respond, you are not smart. When opportunities become active through a trend and yet you cannot make plans, you are not wise. When opportunities emerge through conditions but you cannot act on them, you are not bold." He also said, "Of all avenues of seeing opportunity, none is greater than the unexpected." Disaffection arises from arbitrariness and unfairness, particularly in matters of rewards and punishments, privileges and opportunities. Citing ancient tradition, Zhuge Liang wrote in his advice for commanders, "Do not turn from the loyal and trustworthy because of the artifices of the skilled but treacherous. Do not sit down before your soldiers sit down, do not eat before your soldiers eat. Bear the same cold and heat as your soldiers do; share their toil as well as their ease. Experience sweetness and bitterness just as your soldiers do; take the same risks that they do. Then your soldiers will exert themselves to the utmost, and it will be possible to destroy enemies."

Tian Ji asked, "Once a moving battle line has been established, how does one get the warriors to obey orders without fail when going into action?"

Master Sun said, "Be strict, and indicate how they can profit thereby."

■■ The effort inspired by commonality of purpose is by nature greater, more genuine, and more reliable than the effort inspired by authoritarian demands or fixed wages alone. This point is underscored in the following question and answer.

Tian Ji asked, "Are rewards and punishments critical to warriorship?"

Master Sun said, "No. Rewards are means of encouraging the troops, to make the fighters mindless of death. Punishments are means of correcting disorder, making the people respect authority. These can enhance the odds of winning, but they are not what is most crucial."

■ ■ There are inherent limits to rewards and punishments. Excess in presentation of rewards can be ruinous because of material cost, and by the creation of secondary competition. Excess in punishments can be ruinous because of cost in personnel, and by the creation of an atmosphere of fear and suspicion.

Tian Ji asked, "Are planning, momentum, strategy, and deception critical to warriorship?"

Master Sun answered, "No. Planning is a means of gathering large numbers of people. Momentum is used to ensure that soldiers will fight. Strategy is the means of catching opponents off guard. Deception is a means of thwarting opposition. These can enhance the odds of winning, but they are not what is most crucial."

■ ■ An inspiring plan can magnetize attention and galvanize efforts, but this cannot guarantee positive environmental conditions. Momentum can join a multitude of smaller energies into a stream of major force, but this cannot guarantee the accuracy of aim and direction needed to overcome an intractable obstacle. Strategy can enable one to outwit adversaries when it works, but that cannot prevent them from regrouping and counterattacking. Deception may throw opponents off your trail or off their guard, but that does not guarantee an effective offense to put an end to the conflict. All of these things may have their place in tactical action, in short, but no one of them is sufficient to be in itself quintessential to victory.

Flushed with anger, Tian Ji retorted, "These six things are employed by all experts, and yet you, Maestro, say they are not crucial. If so, then what is crucial?"

Master Sun replied, "Sizing up the opposition, figuring out the danger zones, making sure to survey the terrain, . . . are guiding principles for commanders. To make sure you attack where there is no defense is what is crucial to warriorship. . . ."

▪ ▪ Part of Sun Bin's reply is also missing here, but the overall sense of the passage emphasizes preparedness and surprise. Both of these factors are stressed throughout classical strategic literature.

Tian Ji asked, "Is there a principle according to which a deployed army should not engage in combat?"

Master Sun answered, "Yes. When you occupy a narrow strait and have further increased defensive fortification of this fastness, be quiet, be on the alert, and do not move. Let nothing seduce you, let nothing anger you."

▪ ▪ When you are in a secure position, if you rise to an enemy's bait and let yourself be drawn out through greed or rage, then you give up your security to expose yourself to indefinite risks.

Tian Ji asked Master Sun, "Is there a principle according to which one should not fail to engage in combat even if the opposition is numerous and powerful?"

Master Sun said, "Yes. Fortify your ramparts to enhance determination, solidify group cohesion with strict uprightness. Evade them to make them haughty, lure them to tire them, attack where they are unprepared, act when they least expect it, and make sure you can keep this up."

■ ■ Evading a powerful enemy to give the impression of weakness or lack of confidence is a tactic to induce arrogance, complacency, and overconfidence in order to weaken the enemy's tension and attention. Luring the enemy on fruitless chases is a tactic to wear down the enemy's stamina and patience. Acting outside of expectation and striking where there is no defense are general principles of strategy, but they are particularly recommended in cases where there is so much difference in relative strength that direct confrontation is unfeasible.

Tian Ji asked Master Sun, "What is the Awl Formation for? What is the Goose Formation for? What are elite troops for? What is rapid-fire shooting with powerful bows for? What is a whirlwind battle line for? What are common soldiers for?"

Master Sun replied, "The Awl Formation is for piercing tight defenses and breaking edges. The Goose Formation is for sniping on flanks and responding to changes. . . . Elite troops are for crashing through battle lines to capture commanders. Rapid-fire shooting with strong bows is for ease of battle and the ability to hold out for a long time. A whirlwind battle line is for ... Common soldiers are for sharing the work to bring about victory."

■ ■ The Awl Formation may be thought of as an intense, acute concentration of energy, especially adapted to breaking through obstacles and breaking down resistance. The Goose Formation may be envisioned as still having a concentrated focus, but also maintaining a broader peripheral consciousness and capacity; the expanded scope of its breadth of action is particularly useful for picking off opponents from the side, while the combination of sharply focused and evenly distributed concentration is especially suitable for effective adaptivity in action.

The elite vanguard, supporting artillery, and common soldiers may be translated into civil terms as research, development, and production. Research may be likened to an elite vanguard, which must break through the barriers of existing convention to seize the potential of the unknown. Development supports research by pragmatic follow-up, through which the potential advantages brought to light by research can be tested and proved through transformation into concrete practicalities, so that research continues because of demonstrations of its utility in development of what is useful. Production rationally follows proof of utility, enabling the benefits to be actualized on a public scale.

Master Sun added, "Enlightened rulers and knowledgeable commanders do not expect success by common soldiers alone."

▪ ▪ In both martial and productive endeavors, success is obtained through the cooperation of people of different skills, talents, and capacities. When the potentials inherent in these different capacities are each activated and deployed in such a way as to bring about their maximum collective effect, then they may be said to be cooperating. Cooperation is not simply a matter of everyone doing the same thing regardless of their individual capacities.

To extend this remark of Master Sun to the previous simile of research, development, and production, it may be observed in modern history that economies based on production without research and development have been or become more dependent and more vulnerable than those that have combined research, development, and production within themselves.

It may be, of course, that all economies have, somewhere within them, all three of these elements in some measure. The fact is, however, that in the modern currency/credit-based global economy, research and development can be and have been both forced to and allowed to become more vestigial in many economies at various times than is really healthy— either for the local economy in question or, in the long run, for the global economy itself.

These observations include all tiers of an economy. Ongoing research and development are as important in relation to production in such apparently diverse domains as animal husbandry, agriculture, and the service sector as they are in science, technology, and manufacturing industries.

When Master Sun emerged from these interviews, his disciples asked him about the questions of King Wei and his general, Tian Ji. Master Sun said, "King Wei asked about nine matters, Tian Ji about seven. They are close to knowledge of warriorship, but they have not yet reached the Way.

"I have heard that those who are always trustworthy as a matter of course will flourish, and those who act justly . . . Those who use arms without preparation will be wounded, while armed desperadoes will die. The third generation of Qi is a worry!"

■ ■ Master Sun Bin's fears for Qi were in fact borne out by history in three generations. Although Qi had become one of the most powerful of states in the aftermath of the breakup of the old Zhou dynasty federation, eventually it fell through intrigue and poor judgment. Five smaller states applied to Qi for help against the depredations of the rapacious state of Qin (Ch'in), which eventually would take over all of ancient China and establish the first empire. As it happened, one of the chief advisors of the reigning king of Qi sold out to Qin, accepting bribery to advocate the Qin cause at the Qi court. Siding with Qin on the advice of this traitor, Qi was betrayed and annexed by its supposed ally. The mighty Qin armies then overthrew the five smaller states with ease.

It may seem like sound strategy, if not sound morality, to help the powerful against the weak; and this may have been a reason for the king of Qi to accept and pursue what turned out to be a ruinous tactic recommended by a traitor. Not only did the king of Qi violate the traditional Taoist morality of warfare teaching that the weak should be protected against the strong; from a purely strategic point of view, he also overlooked the tactical potential latent in the very desperation of the five smaller states appealing for assistance against a powerful common enemy.

[*Tian Ji asked Sun Bin, "Would it be effective if my troops in the field keep strengthening their barricades?"

Master Sun replied, "This is the question of an enlightened commander, one which people overlook and do not stress. It is also one detested by opponents."

■ ■ It is one thing to place a task force in the field and expect it to get the job done; it is another to see to the ongoing development and adaptation of this force to emergencies and changing conditions.

Tian Ji said, "Can I hear about it?"

Master Sun answered, "Yes. This is used to respond to sudden changes, or when in confined, closed-off deadly grounds. This is how I captured Pang Juan and Prince Shen."

■ ■ It is strategically necessary to provide for the ability to handle unforeseen developments, unpredicted shifts in the action, and unexpected impasses. The capacity to strengthen defenses under such adverse conditions is essential to mount an effective offense.

Tian Ji said, "Fine, but those events are already past, and I have not seen the formations involved."]

Master Sun said, "Barbed wire can be used to serve the function of a moat, wagons can be used to serve as a

---

*Brackets enclose reconstructions.

barricade. Shields can be used to serve the function of rampart blinds. Long weapons come next, as a means of helping out in danger. Small spears are next, to back up the long weapons. Short weapons are next, to inhibit the enemy's return and strike him when he flags. Bows are next, to serve in place of catapults. The center has no one in it, so it is filled with . . . When the soldiers are set, the rules are fulfilled.

"The code says, 'Place the bows after the barbed wire, then shoot as is proper. Atop the ramparts, bows and spears are half and half.'"

■ ■ A basic principle of practical adaptation is to employ whatever is available at hand to accomplish the task. This includes the art of substituting what one has for what one lacks, and the ability to organize resources in such a way as to maximize the effects obtained from their combination and cooperation.

"A rule says, 'Act after having seen what spies sent out come back and say . . . Keep watchers at intervals where they can be seen. On high ground, have them arrayed rectangularly; on low ground, have them arranged circularly. At night, let them signal with drums; during the day, let them signal with flags.'"

. . .

■ ■ Self-knowledge and knowledge of the opposition are considered critical to successful prosecution of the art of war. The main infrastructures connected with these tasks are a system of external intelligence gathering and a system of internal communications.

Master Sun said, "The victory of a militia lies in its elite corps, its courage lies in order, its skill lies in configuration and momentum, its advantage lies in trust, its effectiveness lies in its guidance, its richness lies in quick return, its strength lies in giving the people rest, its injury lies in repeated battle."

▪ ▪ The elite corps is the vanguard that breaks through or smashes down the edge of an enemy's line of attack or defense, thus creating a loss of momentum or a gap of vulnerability.

Courage is said to be a matter of order in that an orderly formation or organization unites the efforts of people having diverse physical and psychological capabilities, thereby evening out individual disparities; the bold and mettlesome bolster and encourage the less robust, while the presence of weaker and more cautious elements restrains the overly rambunctious.

Skill in configuration and momentum is a matter of organizing people in such a way that they operate as one unit, the force of which can then be directed coherently to achieve an intense focus of power and impact.

Advantage and effectiveness lie in trust and guidance because trust in leadership unifies the people and empowers the leadership. Without guidance, trust is blind; without trust, guidance is powerless.

Richness lies in quick return because this is the way to avoid excess expenditures of constructive energy and material resources. Strength lies in rest because this is the way to avoid useless waste and recover from exertion. Injury lies in repeated battle because continued expenditure of energy and material resources inevitably wears down the strength of a force, even a winning force, ultimately making it vulnerable to loss of capacity to avoid, resist, or withstand antagonistic factors.

Master Sun said, "Acting with integrity is a rich resource for warriors. Trust is a distinguished reward for warriors. Those who despise violence are warriors fit to work for kings. Those who win many cohorts overcome . . ."

■■ Acting with integrity is what wins the trust of leaders, commanders, colleagues, and subordinates, as well as the people at large. Trust solidifies and empowers working relationships, enabling the individual to operate at full personal potential, with the effective cooperation of others.

Warriors who despise violence are fit to work for kings in two important senses. One of the most ancient principles of the art of war is that the best victory is won with the least violence; those who despise violence and yet are warriors are those who are most efficient at their work. Warriors who are fond of violence, furthermore, have a private motivation and cannot be trusted to fight for a public cause; it is those who despise violence who can only be moved to go into battle under conditions of objective necessity. The *Tao Te Ching* says, "Those who enjoy killing cannot get their will of the world." Also, "When you win a war, you celebrate by mourning."

Master Sun said, "There are five conditions that always lead to victory. Those who have authorized command

over a unified power structure are victorious. Those who know the Way are victorious. Those who win many cohorts are victorious. Those whose close associates are in harmony are victorious. Those who take the measure of enemies and size up difficulties are victorious."

▪ ▪ A unified power structure can be expected to be more effective than one that is internally ruptured or fragmented. The Way, according to Sun Bin's predecessor Sun Wu, means "inducing the people to have the same aim as the leadership," thus achieving internal unity of aspiration as well as external unity of organization.

Winning support is naturally conducive to success, but disharmony and lack of integrity within an inner circle of leadership will undermine effectiveness. Knowledge of conditions, of adversaries as well as of critical environmental circumstances, is essential to effective employment of capacities and resources.

Master Sun said, "There are five things that always lead to failure. Inhibiting the commander leads to failure. Not knowing the Way leads to failure. Disobedience to the commander leads to failure. Not using secret agents leads to failure. Not winning many cohorts leads to failure."

▪ ▪ The skills of a directorate cannot materialize in action without an effective organizational structure and chain of command responsive to its initiatives. Inability to achieve this degree of order, by the same token, not only thwarts leadership but is also a failure of the directorate itself. Poor leaders and recalcitrant followers earn each other's mistrust, perhaps because their common cause does not really motivate them, or because their private interests are originally too strong and too disparate to achieve unity of purpose and effort.

The use of secret agents is for the purpose of collecting vital information and the purpose of disseminating crucial misinformation. Secrecy is involved in gathering information because knowledge is power and therefore guarded; secrecy is involved in spreading misinformation to maintain the effect of illusion.

Master Sun said, "Victory lies in consummation of . . . , a clear system of rewards, selecting elite troops, and taking advantage of enemies. . . . This is called the security of a great military."

■ ■ A clear and reliable system of rewards is established to create a unified motivational structure capable of effectively directing the attention and effort of personnel.

Elite troops, whose function and importance as vanguard forces were defined earlier, need to be chosen expertly, based on actual capacity, training, and accomplishment.

The real point of taking advantage of enemies is to win by superior tactical skill rather than by overwhelming violence or force. It is based in the first place, of course, on the premise that the situation has already reached the point where enmity exists and conflict can no longer be avoided by any means.

Master Sun said, "There is no command without leadership.

"[There are three elements of] order: First is trust, second is loyalty, third is willingness. Wherein is loyalty?

Loyalty to the government. Wherein is trust? In reliable rewards. Wherein is willingness? Willingness to get rid of the bad.

"Without loyalty to the government, one may not presume to employ its military. But for reliability in rewards, the peasants will not be virtuous. But for the willingness to get rid of the bad, the peasants will not be respectful."

▪ ▪ A militia, or a special task force of any kind, may accomplish something with its resources and skills, but if the effect is not in harmony with the legitimate underlying aims of the nation or the organization—which include the policies of the rulership or directorate as well as the aspirations of the citizens or the workers—it will be impossible to maintain lasting success and build upon successive achievements.

Master Sun said, "Between sky and earth, nothing is as noble as humanity. . . . The right seasonal timing, the advantages of the terrain, harmony among personnel—if these three things are not gained, there is calamity even in victory. Therefore it is better to give before fighting, only doing battle when there is no choice."

▪ ▪ Even victory is calamitous without the right seasonal timing, the advantages of the terrain, and harmony among personnel, because under these conditions victory will have been won at the cost of loss of productive labor, environmental destruction, and excessively high casualty rates. The Taoist classic *Tao Te Ching* says, "If one were bold but had no mercy, if one were far-ranging but not frugal, if one went ahead without deference, one would die."

The conclusion that it is "better to give before fighting, only doing battle when there is no choice" is also reflected in the *Tao Te Ching*, which claims that this is ancient philosophy: "There are sayings on the use of arms: 'Let us not be aggressors, but defend.' 'Let us not advance an inch, but retreat a foot.' "

"Thus when you have fought for the tranquility of the time, then you do not work the masses anymore. Those who do battle wrongly or unmethodically gain small victories by attrition."

▪ ▪ The special efforts and allocations needed to meet emergencies become ruinous if continued compulsively after the job has been done. The *Tao Te Ching* says, "Calculated sharpness cannot be kept for long. . . . When one's work is accomplished honorably, to retire is the natural way."

The victories of the unjust and unmethodical are attained by attrition because they are gained by fighting when honest and innocent people need to be about their own business.

Master Sun said, "Those who win six out of ten battles go by the stars. Those who win seven out of ten battles go by the sun. Those who win eight of ten battles go by the moon. . . . Those who win ten out of ten battles have skilled commanders yet give rise to calamity. . . ."

■ ■ Those who win all their battles can give rise to calamity by draining resources through continued prosecution of warfare; by creating an aggressive momentum, an appetite for conquest; and by falling prey to complacency and carelessness. There is an ancient saying that repeated victory in repeated warfare produces a haughty leadership commanding an exhausted populace, eventually thereby ruining a nation.

". . . There are five things that make for failure; with even one of these five, you won't win. Thus among ways of war, there are cases where many people are killed but the commanders and troops are not captured, there are cases where commanders and troops are captured but their base camp is not taken, there are cases where a base is taken but the general is not captured, and there are cases where the army is overthrown and the general killed. So, if you find the Way, no one can survive against you."

■ ■ The ancient text is broken, so it is not clear what is intended here by the "five things that make for failure." There appears later in the text, however, an extensive list of failures in commanders.

The sense of the text that does remain is that there are many grades of victory and defeat, many shades of gray. Part of the art of war is understanding how final the outcome of a particular defeat or victory really is, seeing how gains might be lost and how losses might be regained, using this knowledge to plan for security or recovery. Only with comprehensive perspective and fluidly adaptable strategy is it possible to deal unfazed with all sorts of contingencies, even those seeming most desperate.

Master Sun said, "One who leads a militia with inade-quate intelligence is conceited. One who leads a militia with inadequate courage has an inflated ego. One who leads a militia without knowing the Way and does battle repeatedly without being satisfied is surviving on luck."

■■ Unless one has adequate information and also the intellectual ability to process it usefully, one cannot willfully exercise command without an inflated opinion of one's abilities; thus defect is added to lack, providing for a perilous situation. One who takes on leadership in spite of such dangers is foolhardy, not courageous; and one who takes on leadership in psycho-logical compensation for inner lack of fortitude is supremely egotistical, endangering others for personal pride. One who takes on leadership with nothing but witless ambition may get somewhere by dint of perseverance, but no gain attained in this manner can be stabilized safely on a permanent and peaceful basis.

"Bringing security to a large country, expanding a large dominion, and safeguarding a large populace can only be done by knowing the Way. Knowing the Way means knowing the pattern of the climate and the lay of the land, winning the hearts of the people, knowing the con-ditions of enemies, knowing how to set up the eight bat-tle formations, engaging in combat only when it is obvi-ous you will win, otherwise keeping your peace; this is the kind of commander appointed by a successful ruler."

▪ ▪ The secret of the master warrior is knowing when to fight, just as the secret of the artist is knowing when to perform. Knowledge of technical matters and methods is fundamental, but not sufficient to guarantee success; in any art or science of performance and action, direct perception of the potential of the moment is crucial to execution of a master stroke.

Master Sun said, "The use of eight battle formations in combat is based on the advantages of the terrain, using whichever of the eight formations is most suitable. Deploy a battle formation in three parts, each with a vanguard and a backup, each awaiting orders to act, acting only on orders. Use one to fight, two to defend; use one to invade, two to rally.

"When an opponent is weak and confused, send your elite troops in first to take advantage of this. When an opponent is strong and orderly, send your lesser troops in first to lure them.

"When chariots and cavalry are involved, divide them into three groups; one to the left, one to the right, and one in the back. On even ground, use more chariots; in narrow gorges, use more cavalry. On perilous ground, use more archers.

"Whether the ground is rugged or easy, it is imperative to know what ground is viable and what ground is deadly; occupy the viable and attack the deadly."

Master Sun said, "Generally speaking, a course over terrain through sunny ground is called 'outside,' while one through shady ground is called 'inside.'"

▪▪ In metaphorical terms, the "outside" is the obvious, the evident, the open and aboveboard; the "inside" is the subtle, the concealed, the ulterior or underhanded. The "outside" in this sense may also refer to common consensus, the "inside" to private or covert power. The point of defining these distinctions as they apply to a given situation is to match the nature of a feasible approach to fit the character of an accessible route in the process of pursuing chosen aims.

"The straight and direct is called 'rope,' while the crooked and tortuous is called 'string.' When properly organized according to the character of the route, a battle formation does not get confused. Those on a straight way thrive, while those on a tortuous course half die."

▪▪ It may be wondered why anyone would take a tortuous course with this understanding. The answer, aside from real or feigned incompetence, may simply be lack of choice, one of the primary motivations of warriors following the tradition of *The Art of War*.

This aphorism applies to the moral dimension of behavior as well as the strategic aspect. Truth or honesty may seem inconvenient under certain circumstances, but the compensation is freedom from confusion and conservation of energy.

The whole process of creating and maintaining false appearances to conceal and foster ulterior motives requires so much time and energy for its own operation that this preoccupation alone can become a motivation in itself that is powerful enough, however secondary it may be, to turn into a compulsive mode of behavior.

"In general, when it comes to the matter of a battle ground, the sun is the essential element."

■ ■ On an actual battlefield, the most advantageous position to occupy in relation to the sun is to have it at your back and in your opponent's eyes. In metaphorical terms, as the source of light that makes it possible to maneuver, the sun stands for intelligence. Strategically, intelligence means reconnaissance and information as well as the specific means and methods of gathering, processing, and applying knowledge. The question that needs to be considered first is what sources and techniques of intelligence are practical under given conditions.

"Wind may come from eight directions, and must not be forgotten."

■ ■ Depending on its direction in relation to the direction of tactical maneuvers on a battlefield, wind affects vision, hearing, coordination, and stamina. Symbolically, wind is traditionally used to represent external influences that affect states of mind. The "eight winds" are gain and loss, censure and praise, honor and disgrace, pain and pleasure. Insofar as psychological states influence personal interaction and professional performance, the action of the "eight winds" must be considered in the course of organizing and managing a group work situation or developing and implementing an operational strategy of any kind.

"Crossing water, heading up an incline, or going against the current of a river, camping on deadly ground, or facing woods, are equally worthy of note because these are not conducive to victory."

■ ■ Crossing water is perilous because the process of the passage creates inherent vulnerability to attack, difficulty of defense, and inhibition of movement. A maneuver is not conducive to victory if it puts one even temporarily in the position of a "sitting duck" to adversaries, if it requires an excessive expenditure of effort and attention, or if it involves placing oneself in the midst of compromising obstacles under pressure, or even under fire.

By heading up an incline, going the hard way, not only do you lose the advantages of momentum and gravity for movement or offense, you also turn these forces against your own defensive interests.

Going against the current not only saps your strength, it also puts you directly in the firing line of whatever comes down the current from upstream, by chance or by hostile design. Going against the current of affairs not only drains your energy, it places any results of effort beyond the pale of contemporary relevance.

Camping on deadly ground means occupying an indefensible and inescapable position, sitting in an open trap, waiting for someone to shut it. Facing woods is situating yourself in a milieu where malefactors and interlopers can readily conceal themselves in the surroundings.

"Mountains stretching southward are viable mountains, mountains stretching eastward are deadly mountains. Water flowing eastward is viable water, water flowing northward is deadly water. If it does not flow, it is stagnant water.

"The order of superiority of five terrains is as follows: Mountains are superior to high hills, high hills are superior to low hills, low hills are superior to rolling ground, rolling ground is superior to wooded flatlands.

"The five outstanding kinds of vegetation are thickets, brambles, hedges, reeds, and sedges.

"The order of superiority of five kinds of earth is as follows: Green overcomes yellow, yellow overcomes black, black overcomes red, red overcomes white, white overcomes green.

"The five deadly terrains are: natural wells, natural bowls, natural entanglements, natural clefts, and natural pitfalls. These five graveyards are deadly ground, so do not stay there.

"Do not go downhill in spring, do not go uphill in autumn. The main body of the army and the battle formations should not be arrayed to the forward right; they should circle to the right, not the left."

■■ The advantage or disadvantage of a particular element or configuration of a situation depends not only on its own specific characteristics, but also on its interrelationship with other factors and its place in the total context. Factors to examine in making strategic assessments include elements of protection versus vulnerability, concealment versus exposure, freedom of movement versus impediment and restriction, clarity of perspective and vision versus obstruction and partiality, fertility or supportiveness versus aridity or hostility. When the measures of these various factors and their interplay have been assessed, then it is possible to develop a more objective picture of the potential and limitations inherent in a given situation.

Master Sun said, "Fangs and horns, claws and spurs, harmonizing when pleased, fighting when angry—these are in the course of nature and cannot be stopped. Therefore those who have no natural defenses plan strategically for themselves; this is the business of wise leaders.

"The Yellow Emperor invented the sword, symbolizing it by the battle line. Hou Yi invented the bow, symbolizing it by a rush of force. King Yu invented boats and chariots, symbolizing these by adaptive change. King Tang and King Wu invented spears and halberds, symbolizing standard signals. These four are functions of weaponry."

▪ ▪ While traditional Taoist military science condemns militarism as both immoral and inefficient, nevertheless, defensive, protective, peacekeeping, and punitive capabilities are considered rational and natural. Sun Bin follows tradition here in naming the martial actions and contributions of a variety of ancient sage kings and culture heroes to justify the judicious use of arms for pacification and order.

"In what sense is a sword a battle line? You may wear a sword all day without necessarily using it; hence the saying, 'Set out a battle line, but without fighting.' Consider a sword as a battle line: If the sword has no sharp point, even the bravest warrior will not dare . . . ; if a battle line has no elite vanguard, anyone who dares lead it forward without exceptional courage is extremely ignorant of military science. If a sword has no handle, even a skilled warrior cannot go ahead . . . ; if a battle line has no backup, anyone who dares to lead an advance without being a skilled warrior is ignorant of military affairs."

■ ■ A sword must have a point, a cutting edge, a handle, and a ridge. A task force must have direction, skills, maneuverability, and backbone. Direction means the relationship between objective aims and active leadership. Skills need to be deployed selectively, according to conditions, and applied to precise objectives. In order to effect accurate and useful direction of skills, a mechanism of command and control is necessary. In order to carry out directions, an organization needs an adequate degree of inner cohesiveness, such as can be achieved by a commonly shared moral backbone.

In these senses, a sword can be a symbol or a metaphor for a battle line, which can in turn represent a task force of any kind. To wear a sword without necessarily using it means to be prepared but not anxious; the force is not there for its own sake, but for a specific purpose. To operate the force when it is not necessary is a wasteful mistake in itself, and can also evoke undesirable reactions from the political, social, economic, and natural environments.

"Thus when you have a vanguard and a backup with unshakable trust in each other, opponents will flee. If you have neither vanguard nor backup, ..."

■ ■ The vanguard is needed to make the initial cracks in the facade of aggressors; the backup is needed to finish the job of breaking down and through the enemy front. The key expression here is "with unshakable trust in each other." This inner cohesion is the element that fortifies a group to the degree that it can make opponents flee; one of the critical elements of good leadership is evoking and strengthening mutual trust and internal harmony among members of the group.

"In what sense is the bow a rush of force? Shooting from between shoulder and arm, killing people a hundred paces away without their knowing where it is coming from—this is why the bow is said to be a rush of force."

■ ■ A rush of force may be envisioned as a force or movement initiated or "launched" within a relatively small compass that goes on to exert a wide-ranging effect by the force of that initial momentum. In this sense it may be symbolized by the bow and arrow.

"In what sense are boats and chariots adaptive changes? When high, ..."

■ ■ Boats rise and dip with the waves and the tide, chariots travel up and down hills and around curves. These symbolize adapting responsively to changes in circumstances in the course of progress.

"In what sense are spears and halberds signal standards? . . . Signal standards are mostly flags by day and mostly drums by night, used as means of directing the battle."

■ ■ Spears and halberds are models for signal banners because of their frontline position as well as their length and consequent visibility. Sound is used when visual signals are ineffective. Metaphorically, vanguard weaponry representing signal standards reflects the use of outstanding indicators—such as economic indices, technological developments, or progressive sociopolitical adaptations—to evaluate the state of a nation or community.

"These four things are functions of weaponry. Everybody considers them useful, but no one knows the right way to use them.

"Overall, there are four military sciences: battle formation, force, adaptation, and direction. Thorough understanding of these four is a means of destroying powerful enemies and capturing fierce commanders."

■ ■ Battle formation represents the disposition and deployment of resources. Force represents the energy and momentum of an endeavor or a movement. Adaptation represents the capacity to respond effectively to changes. Direction represents the aim and guidance of energy and effort.

Master Sun said, "If you want to know the conditions of a military force, archery is an appropriate model. The arrows are the soldiers, the bow is the commander, and the archer is the ruler."

▪ ▪ The arrows are the power, the bow concentrates and releases the momentum, while the archer takes aim.

"An arrow is tipped with metal and fletched with feathers so that it will be sharp and fly straight. . . . If you organize soldiers so that the rear guard is heavy while the front is light, they may be orderly when arrayed in battle formation, but they will not obey when ordered to charge the enemy. This organization of soldiers is not in accord with the model of the arrow."

▪ ▪ Insufficient force in the vanguard makes it impossible to open up enough of a gap in the resistance to allow a telling follow-through.

"The commander is the bow: if the grip is not right when the bow is drawn, there will be an imbalance of strength and weakness, resulting in disharmony, such that the force imparted by the two ends of the bow will be unequal, and thus the arrows will not hit the target even if they are properly weighted and balanced. If a commander does not harmonize . . . successfully, they will still not overcome the enemy."

■ ■ A commander has to motivate a group of people uniformly enough to get them to operate in harmony. If some are highly enthused while others are cynical and recalcitrant, the energy of the group cannot be focused accurately and released effectively.

"If the arrows are properly weighted and balanced, and the bow draws true and sends arrows with uniform force, yet if the archer is not right, he still won't hit the target. If the soldiers are balanced [and the commander is competent, if the civil leadership is awry,] they still cannot overcome enemies. . . ."

■ ■ Even the best of resources, human and material, however superbly coordinated, cannot consummate a successful operation if the overall aim of the total force and its momentum are off target.

"Thus it is said, 'The way a militia overcomes an opponent is no different from the way an archer hits a target.' This is the way of warfare."

Master Sun said, "The guiding principle for mobilizing warriors and moving people is the balance scale. The balance scale is the means of selecting the wise and choosing the good. Yin and yang are the means of rallying the masses and meeting opponents. When an accurate scale is restacked . . . as long as it is faithful, it is called inexhaustible."

▪▪ The balance scale is used to represent leadership, because a leader must above all be able to weigh and measure, to assess and evaluate all human and environmental factors relevant to an enterprise or an undertaking. The ability to select appropriate personnel for a specific job is a particularly valuable asset in the exercise of leadership.

As for the use of "yin and yang" to rally people and face adversaries, this has a wide range of meanings, based on the broad spectrum of associations of yin and yang. In basic terms relevant to this discussion, yin may have the meaning of self-effacement, docility, or conformity, complemented by yang as self-assertion, initiative, or activity; these refer to harmonizing with allies (yin) and striking out against enemies (yang).

"When articulating direction and establishing a standard of measure, focus only on what is appropriate."

▪▪ It would seem to be a truism to say that focus should be only on what is appropriate, but the idea of inexhaustibility of an accurate scale mentioned in the text above suggests that there is, as the Chinese say, an "eye" in the word "appropriate." The main idea is that what is appropriate depends on the situation and cannot be determined in a dogmatic or peremptory fashion. Thus with the successive arising of new situations and new realities, reexamination of aims and measures is necessary to ensure the maintenance of effective alignment of efforts with actualities.

"Private and public wealth are one. There are those who have too little life and too much money, and there are those who have too little money and too much life: only enlightened rulers and sages recognize them, and thus can keep them in place. When those who die are not bitter, those who are bereft are not resentful."

■ ■ Private and public wealth are one from the point of view of the totality of the economy; the manner in which wealth circulates back and forth between the private and public sectors defines the economy in certain ways, of which enlightened leadership must be aware in order to understand the real and potential effects of programs and policies.

To have too little life and too much money means to have more wealth than can be effectively used under these conditions; to have too little money and too much life means to have more energy or talent than can be constructively employed under these conditions. The wealth of a society that can balance these two extremes does not leak away.

When people die without bitterness and leave no resentment behind them, that means they did the best they could under the conditions in which they lived.

."When there is an abundance of money and goods, things are easy. When things are easy, the people do not attribute the merit to their rulers. . . . Therefore to accumulate wealth for the people is the means whereby you may accumulate wealth yourself; this is how warriors last. . . ."

■ ■ This key idea of Sun Bin is based on traditional philosophy. In his commentary on the classic *I Ching,* or *Book of Changes,* the educator Confucius wrote, "Those above secure their homes by kindness to those below." Also, "Leaders distribute blessings to reach those below them, while avoiding presumption of virtue." According to the later Taoist Masters of Huainan, who compiled a great deal of ancient philosophical and scientific lore, "When people have more than enough, they defer; when they have less than enough, they contend. When people defer, courtesy and justice are born; when they contend, violence and disorder arise."

Master Sun said, ". . . When you know soldiers are trust-worthy, don't let others alienate them. Fight only when you are sure to win, without letting anyone know. In battle, don't forget your flanks, don't . . ."

▪ ▪ Most of this chapter is missing or corrupt.

Even when people are known to be trustworthy, their loy-alty should not be taken for granted. Interlopers may attempt to alienate them, and complacency or arrogance on the part of leadership makes it easier for divisiveness and recalcitrance to take root in the lower echelons.

Fighting only when sure to win is standard wisdom in the philosophy of the art of war transmitted by Sun Wu and Sun Bin. This policy helps to eliminate wasteful heroics and costly wars of attrition, economizing conflict management to the greatest possible degree. Even the *Thirty-Six Strategies,* full as it is of draconian maneuvers, says at the end, "Of the thirty-six strategies, flight is best."

Flanks should not be forgotten, because otherwise you might be outflanked. In general terms, this means that periph-eral awareness should be deliberately maintained along with centrally focused awareness, so that the power of the essential thrust of an effort or undertaking is not undermined by lack or failure of coordinated backup and support measures.

Master Sun said, "When massing troops to assemble armed forces, the thing to do is stimulate energy. When breaking camp and consolidating forces, the thing to do is keep the soldiers orderly and sharpen their energy. When on a border near an enemy, the thing to do is intensify energy. When the day of battle has been set, the thing to do is stabilize energy. On the day of battle, the thing to do is prolong energy.

". . . , thus awing the soldiers of the armed forces, is means of stimulating energy. The general commands . . . , which command is means of sharpening energy. The general then . . . wears simple clothing to encourage the warriors, as a means of intensifying energy. The general gives an order commanding every soldier to muster three days' rations, and the people in the homes of the nation make . . . ; this is a means of stabilizing energy. The general summons his guard and declares, 'Food and drink should not . . .' Thus energy is prolonged."

▪ ▪ The process of stimulating, sharpening, intensifying, stabilizing, and prolonging energy needs to be rationalized so that it can be repeated when necessary. The original meaning of the word *energy* used here in the text includes mental and physical aspects of energy, and both mental and physical momentum are considered critical to the success of an action. The timing of each stage in the process is crucial, so the key to effective leadership is to coordinate the psychological and physical inspiration and readiness of participants in an action with the timing of developments in the unfolding of actual events.

Master Sun said, "Whenever you set out troops, make battle formations efficient, and organize armed forces, when setting up official posts you should do so in a manner appropriate to the individual, indicate ranks by means of insignia, promote and demote to grade people, march in an orderly fashion to . . . , organize soldiers by homeland, delegate authority to those who are leaders in their own localities. Clarify confusion by signal flags and chariots, disseminate orders by means of gongs and drums."

▪ ▪ Setting up official posts in a manner appropriate to the individual means assigning people to duties and responsibilities matching their capacities and talents.

Ranks are indicated by insignia so that organizational order and chain of command can be made clear in an impersonal manner.

Personnel are graded by promotion and demotion to adjust their positions to their abilities and achievements, and to provide a system of rewards and punishments fully integrated into the functional operation of the organization.

Soldiers are organized by homeland for the sake of the inner cohesion of a unit; authority is delegated to local leaders who already have standing in the eyes of their own people.

"To keep soldiers in line, use the method of following tracks. Camps are to be guarded by the strongest men. Overtake armies by means of a continuous line formation;

adjust the formation to contain disorder. Position your army on high ground, use a cloudlike formation for arrow and missile combat. To avoid being surrounded, use a formation like a winding river. To take out a vanguard, shut off the road; when it is on the verge of defeat, circle around. When going to the rescue, put on pressure from outside. In a hectic battle, use mixed lines. Use heavy arms to face a concentrated force, use light arms to face a scattered force. To attack a secured position, use a moving battlement."

■ ■ The method of following tracks means that each successive individual in a line of movement follows in the tracks of the preceding individual. In general representative terms, this means using available forces of internal cohesion to keep a group action focused on collective aims.

Guarding camp does not offer the glamour, excitement, or opportunity for exploit found on the front lines. Untutored thinking might expect the most powerful or most heroic personnel to be strictly elite vanguard material, but the security of the base of operations is essential if the action is to succeed. If a camp is poorly defended, those in the field can be cut off from behind and isolated; they will have no resort in defeat and no backup in victory.

The configuration of an operation, the disposition of resources and personnel, depends on the aim, the terrain and environmental conditions, and the situation and condition of adversaries. This is why it is said that a successful force has no constant configuration.

High ground is preferred because it is easier to command a view of the terrain, and because it puts attacking opponents at a gravitational disadvantage and makes it possible to launch an assault with extra momentum. The same things could also be said of moral high ground, provided the position is authentic and effective, not a mere posture.

A diffuse cloudlike formation is used for arrow and missile attack because it is thereby possible to rain projectiles over a wide area while minimizing casualties under return fire by spreading out rather than clustering.

A formation like a winding river is used to avoid being surrounded, by repeatedly outflanking adversaries and thereby thwarting attempts to encircle your force.

A vanguard is stopped by blocking off its route of advance and then circling around to isolate it and attack from behind.

When going to the rescue, pressure is put on from outside in order to divert the adversary's attention and energy away from the beleaguered party, thus making it easier to secure escape from a difficult situation.

Using mixed lines in a hectic battle means arraying forces so that they are not restricted by their formation but are able to move in any direction, thus being in a position to give and receive support from all sides in the midst of a chaotic fray.

Heavy arms are used against a concentrated force because of its density, and because of the kind of target it affords; intensely focused assault with heavy arms maximizes the power and efficacy of an attack. Light arms are used against a scattered force for the sake of the mobility needed to oppose a relatively diffuse target.

A moving battle line is used against a secured position to take advantage of the limitations imposed on the maneuverability of an occupying force by the requirements of security and defense.

"Use square formations on level ground, use pointed formations when setting out battle lines facing higher ground.

"Use round formations on rugged ground. Use your
forces strategically with alternating aggressiveness and
withdrawal. Against an orderly battle line, use a square
formation with wings; in a more spread-out battle, close
in like a bird's bill shutting. When trapped in rugged territory, open up a way out by outcircling the enemy; on
grass and sand, you have to cut through out in the open.
When you have won in war and yet still maintain troops
in the field, it is to keep the nation on the alert."

■ ■ A square formation is suited to level ground because it is easy
to set out a tight battlefront with a matching backup and
flanks poised to either circle or spread out. Metaphorically
speaking, on a level field of action—when conditions are fair,
being functionally similar or equivalent on all sides—it is appropriate to proceed in a "square" or conventional manner.

A pointed formation is used when facing higher ground because of the need for a sharp edge to resist and break through
the momentum of a downhill charge. An overwhelming force
should not be met with direct resistance, unless the resistance
can be focused so sharply and aimed so adroitly that it does
not absorb the full force but rather splits it apart.

A round formation is used in rugged territory because in a
circle the positions in the front, rear, and flanks can be spaced
in such a way that communication and contact can be maintained in spite of natural barriers, and the formation can expand, contract, or modify its shape as a coherent whole. In
metaphorical terms, emphasis is on "roundness" or strategic
adaptability when conditions are uneven and unfair, because

conventional methods are not sure to work with predictable efficacy in such a situation.

Alternating aggressiveness and withdrawal are used to confuse and mislead opponents; retreat after an assault is a common tactic to draw an adversary into a compromised position. The notorious "hard cop, soft cop" method of interrogation is an application of this principle. The same tactic was used against prisoners of war by communist Chinese agents in Korea.

A square formation with wings is used to outflank and engulf a contained battle line. Against a more scattered force, individuals or squads can be picked off by closing in from two sides, like a bird of prey snatching an animal in its bill.

When on rugged territory, it is necessary to use the difficulty of the terrain to your own advantage, using natural obstacles to help you to outmaneuver adversaries rather than letting them keep you trapped. Where the ground is flat and open, in contrast, it is necessary to cut right through because there is no natural cover. Here again there is a metaphorical contrast between the use of subterfuge and deviousness when at an unfair disadvantage and the use of a more open and direct approach on an even ground where no one has an unfair advantage.

Maintaining troops in the field after a war is won is normally not indicated because of the drain on the economy. It is only justified tactically when the situation has not been completely stabilized and it is imperative to keep on the alert.

". . . In thick undergrowth, move like a snake; to make it easy when weary, travel in a Goose Formation. In dangerous straits, use a medley of weaponry; when retreating, dissolve into the underbrush."

■ ■ Zigzagging through obstacles like a snake rather than plowing through them like a bulldozer has the advantages of conserving energy, minimizing environmental destruction, and leaving a less obvious trail. Traveling in a Goose Formation makes it easy when weary by positioning people where they can easily keep in contact with each other and come to each other's assistance and yet not stumble over one another. A medley of weaponry is used in dangerous straits because different weapons have different effective ranges and usages, so having a variety of arms at hand increases resources and enhances adaptability. Dissolving into the underbrush when retreating means relying on concealment rather than speed of flight, which is naturally compromised by the fatigue and stress of battle.

"When circling mountains and forests, use circuitous routes and go by stages; to attack cities, use their waterways. Organize night retreats by memo; use relay signals for night alarms. Use talented warriors for double agents. Place troops armed with close-range weaponry where convoys are sure to pass."

■ ■ One may take a circuitous route to outflank an opponent's position, or to weary an opponent in pursuit. The purpose of going by stages is to avoid debilitating weariness. Waterways are convenient for attacking cities because their functional relationship to cities makes them ideal delivery systems for assault forces; waterways can also be blocked or poisoned. Night retreats are organized by memo for security reasons, so that the orders and plans for retreat do not leak out. Relay signals are used for night alarms to compensate for limited visibility.

Talented warriors are used for double agents because their talents can win them the confidence of adversaries for whom they appear to be working. Ambushing convoys reduces defensive expenditures by diminishing the enemy's fighting capacity with minimal effort.

"For incendiary warfare, deliver the fuel in wagons. When setting out a battlefront of blades, use a pointed formation. When you have few soldiers, deploy them with a combination of weapons; a combination of weaponry is a way to prevent being surrounded."

■ ■ The best firepower delivery system to use in a given situation depends on the nature of the particular form of firepower to be employed and the local and temporal environmental factors affecting transport.

The use of a pointed formation in a battlefront of bladed weapons is recommended to maximize the effective range of the weaponry while minimizing the dangers of accidental injuries in a crush.

Deploying a small force with a variety of weaponry is a way to enhance the efficiency and adaptability of each individual warrior. In particular, arming everyone with both long- and short-range weapons increases their chances of preventing a larger enemy from surrounding them by holding the enemy at bay or breaking through attempts to outflank and encircle them.

"Patching up the lines and linking fragments is a way to solidify battle formations. Swirling and interlacing is a

means of dealing with emergencies. A whirlwind kicking up dust can be used to take advantage of unclarity. Hiding out and hatching schemes is a way of provoking a fight. Creeping like a dragon and positioning ambushers is a way to fight in the mountains. . . ."

▪ ▪ Patching lines and linking fragments means regrouping your forces. A mediocre commander, or a mediocre force, is one that does this only after being routed; the true warrior, in contrast, is constantly solidifying in this way, grooming power under all conditions. One of the reasons Turkish captives in North Korea were able to resist communist Chinese brainwashing techniques was because they continually regrouped in spite of all efforts by their captors to destroy group cohesion and leadership.

Swirling and interlacing work together as a way of meeting emergencies. Swirling is a technique of dodging direct onslaughts while simultaneously launching one's own assault from constantly changing angles; interlacing reinforces the swirling lines without compromising their fluidity.

A whirlwind kicking up dust, or mass distraction or confusion of any sort, can be used as a cover for covert operations or sneak attack. It is widely used by thieves, especially pickpockets.

Hiding out and hatching schemes is a way of provoking fights because it arouses the suspicions and fears of enemies. For this reason, the appearance of being open and aboveboard is also used as a diversionary tactic, presenting adversaries with a nonsuspicious front while plotting against them under the cover of their own false sense of security.

Guerrilla tactics are recommended for mountain warfare because of the inherent difficulties of mountainous terrain.

Once in such a situation, the most practical way out is to use the difficulties to one's advantage against adversaries.

"Sneaking up unexpectedly on soldiers is a way of fighting in the dark. Taking a stand on the opposite side of a river is a way of clashing with a smaller force."

▪ ▪ Sneaking up on enemies unawares is more precise, efficient, and economical than random bombardment. Forcing an opponent to cross a natural barrier, rendering itself vulnerable to attack as it does so, is a way of keeping expenditures to an absolute minimum.

"Tattering banners is a way to fool enemies. A chariot train in swift formation is a way to pursue remnant forces."

▪ ▪ Tattering banners means giving the appearance of fatigue and distress in order to make opponents contemptuous, haughty, and therefore careless and unprepared for a hard fight.

More ancient work on strategy draws limits to the distances to which a fleeing enemy should be pursued by foot soldiers and by cavaliers. The reason for this is to limit expenditure of time and energy, and to avoid being lured into ambushes. The use of a chariot train to chase down remnant forces provides for greater swiftness and stamina in advance and withdrawal than can be achieved on foot, and more powerful defensive and offensive capabilities than horseback fighters alone.

"Ability to move an army at a moment's notice is a way to be prepared against those who are stronger. Spreading out over water or swampland is a way to fight with fire."

■ ■ When strength is overbearing, it can be neutralized by yielding; the flexibility to change at an impasse is one variety of this maneuver. Lao-tzu said, "The softest can drive the hardest."

The use of environmental or other ambient factors inherently antagonistic to specific kinds of force is another mode of softness overcoming hardness; here this manner of defense is typified by using water to control fire.

"Retreating under cover of darkness, like a cicada leaving its shell, is a means of luring an enemy on. A light, mobile task force of specially trained troops is used to oppose a blitz attack."

■ ■ The image of the cicada leaving its shell is a traditional representation of strategic maneuvering whereby a semblance or facade is left in place to convey a misleading impression, while the real power or force has been moved elsewhere, poised for a surprise assault on the opponent who has been deceived by appearances.

In the case of a blitz attack, the nature of the action makes it inherently costly to mount direct opposition. Mobility is therefore essential to counter such an attack, so that the most dangerous and destructive waste of a head-on collision may be avoided while more patient and more effective defensive measures are arranged and carried out by strategically harrying, diverting, and splitting up the oncoming force.

"A stiffened and thickened battle line is used to attack fortifications. Making breaks in surrounding ground cover is a way to create confusion."

▪▪ The precise manner in which a battle line is stiffened and thickened with extra weaponry and personnel depends on the characteristics of the fortifications under siege. The general idea is to provide for the flexibility to concentrate or dilute manpower and firepower freely enough to adapt successfully to rapidly changing needs and challenges.

Ground cover providing camouflage under which to maneuver is undoubtedly useful, but unmitigated cover may frustrate an opponent so much as to incite random fire or blanket fire. When breaks are made in the ground cover, in contrast, tactical movements of troops through these breaks can be staged so as to create concrete but false impressions of the strength and disposition of those under cover. Thus instead of the risk of an uncontrolled release of fear, it is possible to take advantage of a calculated manipulation of apprehension.

"Pretending to leave behind a small loss is a way to bait an opponent. Heavy weaponry and severe violence are used in active combat. To maneuver at night, use signals opposite to those used during the day."

▪▪ Appearing to make a concession in order to bait an adversary is a tactic that may be useful when trying to lure an enemy out of a fastness, or when trying to slow down an advancing force without putting up direct opposition.

It seems redundant to say that heavy weaponry and severe violence are used in active combat, but this is an indirect way of teaching, somewhat like making a noise to produce an echo. The point of making such an apparently obvious statement is to emphasize the basic tactical principle that combat is a last resort, that it is better to win by strategy than by violence.

The reason for varying signals is to make them more difficult for the enemy to read. This is an example of the principle that "the unconventional becomes conventional, the conventional becomes unconventional." Surprise tactics and secret usages become routine if they are employed too much; routines have to be changed if the element of surprise is to be exploited.

"Excellent salaries and useful supplies are means of facilitating victory. Firm and strong warriors are needed to repel assaults. . . ."

■ ■ Excellent salaries are means of facilitating success when they are used to attract and maintain superior personnel and dependable loyalty.

The usefulness of useful supplies is another self-evident tautism used as a sound to produce an echo. In this case, the echo is the idea that the utility of supplies is not only a matter of quantity, but also of quality. The question of useful qualities is one that changes according to situations, so every operation needs to be considered in terms of its particular needs.

Firmness and strength are qualities proper to all warriors. The point of saying that such warriors are needed to repel assaults is another way of expressing the principle that these qualities are not properly used for aggression but for defense and prevention.

▪ ▪ [There are significant lacunae in every sentence of this chapter, such as to make it impractical to attempt to produce an accurate and meaningful translation.]

Generally speaking, there are ten kinds of battle formations. There are square formations, round formations, sparse formations, dense formations, pointed formations, formations like a flock of geese, hooklike formations, confusing formations, fire formations, and water formations. Each of these has its uses.

Square formations are for cutting off, round formations are for massing solidly. Sparse formations are for bristling, dense formations are for being impossible to take. Pointed formations are for cutting through, formations like goose flocks are for handling barrages. Hooklike formations make it possible to adapt and change plans, confusing formations are for deceiving armies and muddling them. Fire formations are used for rapid destruction, water formations are used for both offense and defense.

The rule for square formations is to make the center thin and the sides thick, with the main line at the back. The sparse array in the center is used for bristling.

▪▪ "Bristling" refers to giving the illusion of being bigger and stronger than one really is, just as an animal bristles when faced with a natural enemy.

The rule for sparse formations is for added strength and firmness in cases where there is little armor and few people. The warrior's technique is to set up banners and flags to give the appearance that there are people there. Therefore they are arrayed sparsely, with space in between, increasing the banners and insignia, with sharpened blades ready at the flanks. They should be at sufficient distance to avoid stumbling over each other, yet arrayed densely enough that they cannot be surrounded; this is a matter for caution. The chariots are not to gallop, the foot soldiers are not to run. The general rule for sparse formations is in making numerous small groups, which may advance or retreat, may strike or defend, may intimidate enemies or may ambush them when they wear down. In this way a sparse formation can successfully take an elite corps.

The rule for a dense formation is not to space the troops too far apart; have them travel at close quarters, massing the blades yet giving enough room to wield them freely, front and rear protecting each other. . . . If the troops are frightened, settle them down. Do not pursue opponents in flight, do not try to stop them from coming; either strike them on a circuitous route, or break down their elite troops. Make your formation tightly woven, so there are no gaps; when you withdraw, do so under cover. In this way, a dense formation cannot be broken down.

A pointed formation is like a sword: if the tip is not sharp, it will not penetrate; if the edge is not thin, it will not cut; if the base is not thick, it cannot be deployed on the battlefront. Therefore the tip must be sharp, the edge must be thin, and the main body must be thick; then a pointed formation can be used for cutting through.

. . .

In a hooked formation, the front lines should be straight, while the left and right flanks are hooked. With gongs, drums, and pipes at the ready, and flags prepared, the troops should know their own signal and flag. . . .

A confusing formation must use a lot of flags and insignia, and drum up a racket. If the soldiers are in a commotion, then settle them down; if the chariots are disorderly, then line them up. When all is in order, the battle line moves with a shocking commotion, as though it had come down from the sky or emerged from the earth. The foot soldiers come on unstoppably, continuing all day long inexhaustibly.

The rules for incendiary warfare are as follows. Once moats and ramparts have been made, construct another moat. Pile kindling every five paces, making sure the piles are placed at even intervals. A few men are needed to set the fires; they must be fast and efficient. Avoid being downwind; if the fire has overwhelmed you, you cannot fight a winning battle, and you will lose whether you stay put or go into action.

The rule for incendiary warfare is that the ground should be low and grassy, so that enemy soldiers have no way out. Under these conditions, it is feasible to use fire. If it is windy, if there is plenty of natural fuel, if kindling has been piled up, and if the enemy encampment is not carefully guarded, then a fire attack is feasible. Throw them into confusion with fire, shower them with arrows, drum and yell to encourage your soldiers, using momentum to help them. These are the principles of incendiary warfare.

The rule for amphibious warfare is to have a lot of infantry and few chariots. Have them fully equipped so that they can keep up when advancing and do not bunch up when withdrawing. To avoid bunching up, go with the current; make the enemy soldiers into targets.

The rule for warfare on the water is to use light boats to guide the way, use speedboats for messengers. If the enemy retreats, pursue; if the enemy approaches, close in. Be careful about advancing and withdrawing in an orderly manner, according to what is prudent under prevailing conditions. Be on the alert as they shift positions, attack them as they set up a front, split them up as they organize. As the soldiers have a variety of weapons and chariots, and have both mounted troops and infantry, it is essential to find out their quantities. Attack their boats, blockade the fords, and inform your people when the troops are coming. These are the rules of amphibious combat.

▪ ▪ In this chapter, classical tactics are defined according to their usefulness in given situations. All of these strategies are to be found in *The Art of War,* the earlier manual by Sun Bin's distinguished predecessor, Sun Wu. Typical examples include feigning flight to split up opponents and set them up for a counterattack; dividing and regrouping to confuse and overwhelm enemies; seeking the advantage of the terrain according to conditions; feinting to mislead opponents and create openings; attacking where there is no defense; inducing laziness and arrogance in adversaries by appearing irresolute; seducing opponents into ambushes; striking unexpectedly with such speed that there is no time to mount a defense.

Someone asked, "Suppose two armies are facing off with equal fodder and food, comparable personnel and weaponry, both aggressor and defender wary. If the enemy uses a round battle formation for security, how should we attack it?"

[Master Sun Bin replied,] "To strike an opponent like this, divide your forces into four or five groups, one of which closes in and then feigns defeat and flight to give the appearance of fear. Once the opponents see you to be afraid, they will unthinkingly split up to give chase. Thus their security will be disrupted. Now mobilize your cavalry and drummers, attacking with all five groups at once. When your five divisions get there together, all of your forces will cooperate profitably. This is the way to strike a round formation."

"Suppose two armies are facing off, and our opponents are richer, more numerous, and more powerful than we are. If they come in a square formation, how do we strike them?"

"To strike such a force, [using a sparse] formation to [assault] them, contrive to split them up. Clash with them, then appear to run away beaten, then come kill them from behind without letting them know what is going on. This is the way to strike a square formation."

"When two armies face off, suppose the enemy is numerous and powerful, forcefully swift and unyielding, waiting with a battle line of crack troops; how do we strike them?"

"To attack them, it is necessary to divide into three. One group stretches out horizontally. The second group . . . so that the enemy leaders are afraid and their troops are confused. Once both lower and upper echelons are in disarray, the whole army is routed. This is how to strike a battle formation of elite troops."

"When two armies face off, suppose the enemy is numerous and powerful, and stretches out in a horizontal battle line; meanwhile, we set out our front to await them, but we have few troops, and even these are unskilled. How do we strike?"

"You must divide your troops into three battalions. Train a suicide squad; have two battalions stretch out a battlefront, extending the flanks, while the elite specially trained group attacks the enemy's strategic points. This is the way to kill commanders and crash horizontal battle fronts."

"When two armies face off, suppose we have a lot of infantry but ten times fewer chariots than the enemy; how do we strike?"

"Keep to rugged terrain, carefully avoiding wide-open level ground. Level ground is advantageous for chariots, rugged terrain is advantageous for infantry. This is the way to attack chariots."

"When two armies face off, and our side has plenty of chariots and cavalry, but the enemy has ten times as much personnel and weapons as we have, how do we attack them?"

"To attack them, be careful to avoid constricting land formations; induce them to pass through to level, open ground, where your chariots will have an advantage and be able to strike even if the enemy has ten times the men and weaponry you have. This is the way to attack infantry."

"When two armies face off, suppose our supplies are irregular and our personnel and weapons are inadequate, so we have to make an all-out attack on an enemy ten times our size, how do we strike?"

"To strike in this case, once the enemy has occupied a fastness, you . . . turn around and attack where they have no strength. This is a strategy for aggressive contention. . . ."

"When two armies face off, suppose the enemy commander is brave and cannot be intimidated, the enemy's weaponry is powerful, their troops are numerous, and they are in a secure position. Their soldiers are all brave and unruffled, their commander is fierce, their weaponry is powerful, their officers are strong and their supplies are regular, so that none of the local leaders can stand up to them. How do we strike them?"

"To strike in this case, let them think you lack resolve, feign lack of ability, and appear to have a defeatist attitude, so as to seduce them into arrogance and laziness, making sure they do not recognize the real facts. Then, on this basis, strike where they are unprepared, attack where they are not defending, pressure those who have slacked off, and attack those who are uncertain or confused. As long as they are haughty and warlike, when the armies break camp the front and rear battalions will not look out for each other; so if you strike them precisely at this point, it will be as though you had a lot of manpower. This is the way to strike a large and powerful force."

"When two armies face off, suppose the enemy holds the mountains and occupies the defiles, so that we cannot get to them if we are far off yet have nowhere nearby to take a stand; how do we strike them?"

"To strike in this case, since the enemy has withdrawn to a fastness . . . then put them in danger; attack where they are sure to go to the rescue, so as to get them to leave their fastness, and thus find out their intentions; set out ambushers, provide reinforcements to back them up, and strike the enemy troops while they are on the move. This is the way to attack an opponent occupying a fastness."

"When two armies face off, aggressor and defender both arrayed in battle lines, suppose the enemy takes a basketlike formation, so it seems they want us to fall into a trap; how do we strike them?"

"To strike in this case, move so quickly that the thirsty haven't time to drink and the hungry haven't time to eat; use two thirds of your forces, and aim for a critical target. Once they . . . have your best and most well trained soldiers attack their flanks. . . . Their whole army will be routed. This is the way to strike at a basket formation."

▪▪ [This chapter is so fragmentary that even the order of the strips is uncertain.]

In warfare, there is an aggressive party and a defensive party. Aggression requires more troops than defense; when there are twice as many aggressors as defenders, it is still possible to oppose them.

The defender is the one who is first to get set up, the aggressor is the one who is last to get set up. The defender secures the ground and settles his forces to await the aggressor, who comes through narrow passes. . . .

▪ ▪ The terms *aggressor* and *defender* here are not defined in reference to invasion and defense of the homeland of one of the parties by another, but in reference to confrontation on mutually contested ground. The first to get set up is the defender, in terms of defending a claim or a conquest; while the last to get set up is the aggressor, in terms of challenging that claim or conquest.

When soldiers retreat even in face of the threat of decapitation, and refuse to oppose the enemy as they advance, what is the reason? It is because the configuration of forces is unfavorable and the lay of the land is not advantageous. If the configuration of forces is favorable and the lay of the land is advantageous, people will advance on their own; otherwise, they will retreat on their own. Those who are called skilled warriors are those who take advantage of configurations of forces and the lay of the land.

■ ■ The point of these statements, which may seem repetitive truisms, is that authoritarian coerciveness is not ultimately effective, whether in war or in peace, if for no other reason than that there will always be people who follow natural intelligence whatever others may say. True leaders are not those who force others to follow them, but those who are able to harmonize the wills of others and unify the overall direction of their energies.

If you keep a standing army of 100,000 troops, they won't have enough to eat even if the populace has surpluses. . . . There will be more soldiers in camp than in action, and those in camp will have plenty while those in action will not have enough.

■ ■ Standing armies were a comparatively recent development in the time of Sun Bin, but here it is evident that the civil and military economic pressures and imbalances resulting from such a system were quite apparent to him.

If you have an army of 100,000 troops and send them out in battalions of 1,000, the enemy may repel you with battalions of 10,000 each. So those skilled at warfare are skilled at trimming enemies down and cutting their forces apart, like a butcher dismembering a carcass.

Those who are able to split up others' armies and control others' forces are adequate even with the smallest quantity; those who are unable to split up others' armies and control others' forces are inadequate even if they have several times the firepower.

■■ Taking on too much at once can sap any amount of energy and thwart the successful completion of any undertaking. Parceling tasks into manageable portions without losing sight of the overall design of the whole endeavor is one of the arts of leadership at all levels, from personal self-management to corporate, community, and political domains of action.

Do you suppose that the side with the most troops wins? Then it is just a matter of going into battle based on head count. Do you suppose the wealthier side wins? Then it is just a matter of going into battle based on measurement of grain. Do you think the side with sharper weapons and stronger armor wins? Then it would be easy to determine the victor.

Therefore the rich are not necessarily secure, the poor are not necessarily insecure, the majority do not necessarily prevail, minorities do not necessarily fail. That which determines who will win and who will lose, who is secure and who is in peril, is their science, their Way.

■■ According to *The Art of War* by Sun Bin's distinguished predecessor Sun Wu, the "Way" is that whereby the wills of those above and those below are united. In other words, the Way is the guiding ideal, principle, or means of accomplishing collective goals, that which subtends the order and morale of an organization. Without this cohesion, the superiority of numbers, supplies, or equipment cannot guarantee success.

If you are outnumbered by opponents but are able to split them up so they cannot help others . . . the stoutness of their armor and efficiency of their weapons cannot assure them strength, and even soldiers having courage and power cannot use them to guard their commanders, then there is a way to win.

■ ■ Conversely, if it is possible to undermine the cohesion of a more powerful opponent, it is thereby possible to compensate for disadvantages of numbers, arms, or other formal and material factors.

Therefore intelligent governments and commanders with knowledge of military science must prepare first; then they can achieve success before fighting, so that they do not lose a successful accomplishment possible after fighting. Therefore, when warriors go out successfully and come back unhurt, they understand the art of war.

. . .

■ ■ In making preparations for struggle, it is not only necessary to consider how best to prevail, but also how best to handle the aftermath of struggle, how to safeguard the fruits of victory, and how to make the best of further opportunities that arise as a result of success. It is also imperative, of course, to include due consideration of problems, difficulties, and the chances of defeat, in order to be able to "go out successfully and come back unhurt."

Even though an enemy army has many troops, an expert can split them up so that they cannot help each other while being attacked.

Therefore the depth of your moats and the height of your ramparts do not make you secure, the strength of your chariots and the effectiveness of your weaponry do not make you awesome, and the bravery and strength of your soldiers do not make you powerful.

Therefore experts take control of mountain passes and take account of obstacles; they take care of their troops, and are able to contract and expand fluidly. If enemies have many troops, experts can make them as if few; if enemy stores of food are enough to fill their troops, experts can make them starve; if enemies stay in their places unmoving, experts can make them tire. If enemies have won the world, experts can cause division; if enemy armies are harmonious, experts can break them up.

■ ■ Following on the preceding chapter, this one begins by emphatically restating the critical importance of group cohesion, beyond even that of sheer material and energetic factors. One of the essential elements of cohesion, furthermore, is a comprehensive and coherent strategy that can outwardly adapt to all situations while inwardly maintaining integrity of purpose and morale in pursuing goals.

So military operations have four routes and five movements. Advance is a route, and withdrawal is a route; to the left is a route, and to the right is a route. To go forward is a movement, to retreat is a movement; to go to the left is a movement, and to go to the right is a movement. To stay put silently is also a movement.

Experts make sure to master the four routes and five movements. Therefore when they advance, they cannot be headed off in front; and when they withdraw, they cannot be cut off behind. When they go to the left or right, they cannot be trapped on treacherous ground. When they stay put silently, they are not troubled by opponents.

Thus experts drive their enemies to their wits' end in all four routes and five movements. When enemies advance, experts press them in front; when enemies retreat, experts cut them off from behind. When enemies move left or right, experts trap them in rough terrain. When enemies silently stay put, their troops cannot escape trouble. Experts can make enemies put aside their heavy armor and rush long distances by forced double marches, so they cannot rest when they get tired and sick, and cannot eat and drink when they get hungry and thirsty. Thus do experts press enemies to ensure that they cannot win at war.

You eat to your fill and wait for the enemy to starve; you stay put comfortably and wait for the enemy to tire; you keep perfectly still and wait for the enemy to stir. Thus will the people be seen to advance without retreating, tread on naked blades without turning on their heels.

■ ■ Relentless pressure is one way to thwart an opponent's strategy at every step and thereby systematically undermine morale. Made from a position of relative security, unremitting pressure is supported and strengthened by the specific psychological effects visited upon both parties by this sort of tactic.

There are five descriptions of military forces. The first is called awesome and powerful. The second is called proud and arrogant. The third is called adamant to the extreme. The fourth is called greedy and suspicious. The fifth is called slow and yielding.

An awesome and powerful force you treat with humility and softness. A proud and arrogant force you keep waiting with courteous respect. An extremely adamant force you take by seduction. A greedy and suspicious force you press in front, harass at the sides, and use deep moats and high barricades to make it hard for them to keep supplied. A slow and yielding force you terrorize by harassment; shake them up, surround them, and strike them if they come out. If they do not come out, then encircle them.

Military actions have five courtesies and five harsh actions. What are the five courtesies? If it invades a territory and is too courteous, a militia loses its normal state. If it invades a second time and is too courteous, a militia will have no fodder. If it invades a third time and is too courteous, a militia will lose its equipment. If it invades a fourth time and is too courteous, a militia will have no food. If it invades a fifth time and is too courteous, a militia will not accomplish its business.

Violently invading a territory once is called aggression. Violently invading a second time is called vanity. A third violent invasion, and the natives will be terrorized.

A fourth violent invasion, and the soldiers will be given misinformation. A fifth violent invasion, and the militia will be worn out.

Therefore, courtesy and harshness must be intermixed.

■ ■ An invasion or a takeover has to command respect and collaboration without causing terror and disaffection if it is to avoid either absorption and vitiation of its power on the one hand, or resistance and repulsion on the other.

If you want to use unrest among the people of an enemy state . . . to inhibit the strengths of the enemy state's military, you will wear out your own military.

■ ■ Fanning flames of unrest among a people is one way to attack their government and also inhibit the strength of their military by preoccupation with civil disturbance. This does not guarantee, however, that people aroused by such provocations will necessarily side with your cause. This tactic is thus as likely to result in an overall increase in resistance to outside control, thus wearing down the mechanisms by which the attempt to assert control is made.

If you want to strengthen and increase what your state lacks in response to the abundance of an enemy state, this will quickly frustrate your army.

■ ■ Competing with a rival on a sheerly quantitative basis leads to excessively narrow funneling of energy and resources along lines determined too rigidly by fixation on fear of the competition. This results in frustration through lack of flexibility, foresight, and discretionary resources needed to adapt to changing circumstances in the environment at large.

If your preparations are all set, and yet you cannot thwart the enemy's equipment, your army will be disrespected. If your equipment is not effective, while your enemy is well prepared, your army will be crushed. . . .

■ ■ The unspoken point of these apparent truisms is that when objective assessments indicate that you are in such a position, it is better to avoid engagement with the opponent. This is not simply because of the immediate likelihood of defeat, but because of the long-term strategic disadvantages of humiliation and demoralization.

If you are skilled at arraying battle lines, and you know the odds for and against, and know the lay of the land, and yet your army is thwarted time and again, that means you do not understand both diplomatic victory and military victory.

■ ■ Purely military or strategic factors are not considered sufficient guarantees of victory. This is why the Way, which in this context means the social rationale for action, the moral/morale factor, is regarded so critically even in what would otherwise seem to be strictly tactical matters.

. . . [If] the armed forces are incapable of great success, that means they do not recognize appropriate opportunities. If the military loses the people, that means it is unaware of its own faults and excesses. If the armed forces require much effort to accomplish little, that means they do not know the right timing. If the military cannot overcome major problems, it is because it cannot unite the hearts of the people. When the armed forces have a lot of regrets, it is because they believed in what was dubious. When warriors cannot see fortune and disaster before these have taken shape, they do not know how to prepare.

■ ■ Disorientation, disaffection, inefficiency, disunity, delusion, lack of foresight—these are basic problems that undermine successful collective effort. Understanding why they happen is as important as recognizing them when they happen.

If warriors are lazy when they see good to be done, are doubtful when the right time to act arrives, get rid of wrongs but cannot keep this up, that is the way to stagnation. When they are honest and decent even though ambitious, polite even when favored, strong though yielding, flexible yet firm, this is the way to thrive.

If you travel the path to stagnation, even heaven and earth cannot make you flourish. If you practice the way to thrive, even heaven and earth cannot make you perish.

. . .

■ ■ An ancient Taoist saying goes, "My fate depends on me, not on Heaven." Strategists did not believe in predestination and did not encourage people to consult fortune-telling books and hope for the best. They taught people to examine their own situations and their own actions, and to take conscious responsibility for their own behavior and its consequences.

Commanders must be just; if they are not just, they will lack dignity. If they lack dignity, they will lack charisma; and if they lack charisma, their soldiers will not face death for them. Therefore justice is the head of warriorship.

■ ■ Justice also means duty. Commanders who are not just and do not command justice lack dignity because those under their command will not fear to be unruly.

Commanders must be humane; if they are not humane, their forces will not be effective. If their forces are not effective, they will not achieve anything. Therefore humaneness is the gut of warriorship.

■ ■ If commanders are not humane, their forces will not be effective because there will be no bond of loyalty between them. The troops of a leader who is not humane will lack motivation to fight loyally for the cause.

Commanders must have integrity; without integrity, they have no power. If they have no power, they cannot bring out the best in their armies. Therefore integrity is the hand of warriorship.

■ ■ Without integrity, commanders have no power because they do not back up their words with their deeds and therefore cannot inspire confidence and trust.

Commanders must be trustworthy; if they are not trustworthy, their orders will not be carried out. If their orders are not carried out, then forces will not be unified. If the armed forces are not unified, they will not be successful. Therefore trustworthiness is the foot of warriorship.

■ ■ Trustworthiness cements the relationship of commander and forces, letting the forces know they can expect to be rewarded for doing well and punished for cowardice or unruliness.

Commanders must be superior in intelligence; if they are not superior in intelligence, . . . their forces lack [resolution]. Therefore resolution is the tail of warriorship.

■ ■ Resolution derives from intelligence through the repose of confidence in an intelligent plan of action.

▪ ▪ [This chapter is all in fragments. The next five chapters have some lacunae, but they are largely descriptive and self-explanatory. They need no elucidation, but will nevertheless yield more to reflection.]

These are failings in commanders:

1 They consider themselves capable of what they are unable to do.
2 They are arrogant.
3 They are ambitious for rank.
4 They are greedy for wealth.
5 . . .
6 They are impulsive.
7 They are slow.
8 They lack bravery.
9 They are brave but weak.
10 They lack trustworthiness.
11 . . .
12 . . .
13 . . .
14 They lack resolution.
15 They are lax.
16 They are lazy.
17 . . .
18 They are vicious.
19 They are self-centered.
20 They are personally disorderly.

Those with many failings suffer many losses.

These are losses of commanders:

1 When they lose purpose in their maneuvering, they can be beaten.

2 If they take in unruly people and deploy them, keep defeated soldiers and put them back in battle, and presume to have qualifications they really lack, they can be beaten.

3 If they keep arguing over judgments of right and wrong, and keep debating over elements of strategy, they can be beaten.

4 If their orders are not carried out and their troops are not unified, they can be beaten.

5 If their subordinates are refractory and their troops won't work for them, they can be beaten.

6 If the populace is embittered against their armed forces, they can be beaten.

7 If an army is out in the field too long, it can be beaten.

8 If an army has reservations, it can be beaten.

9 If the soldiers flee, they can be beaten.

10 . . .

11 If the troops panic repeatedly, they can be beaten.

12 If the course of a military operation turns into a quagmire and everyone is miserable, they can be beaten.

13 If the troops are exhausted in the process of building fortifications, they can be beaten.

14 . . .

15 If the day is coming to an end when there is yet far to go and the troops are eager to get there, they can be beaten.

16 . . .

17 . . . the troops are afraid, they can be beaten.

18 If orders are repeatedly modified and the troops are dilatory, they can be beaten.

19 If there is no esprit de corps and the troops do not credit their commanders and officers with ability, they can be beaten.

20 If there is a lot of favoritism and the troops are lazy, they can be beaten.

21 If there is a lot of suspicion and the troops are in doubt, they can be beaten.

22 If commanders hate to hear it when they've erred, they can be beaten.

23 If they appoint incompetents, they can be beaten.

24 If they keep their troops out in the field so long as to undermine their will, they can be beaten.

25 If they are scheduled to go into combat but their minds are still divided, they can be beaten.

26 If they count on the other side losing heart, they can be beaten.

27 If their actions hurt people and they rely on ambush and deception, they can be beaten.

28 . . .

29 If the commanders oppress the soldiers, so the troops hate them, they can be beaten.

30 If they cannot get out of narrow straits in complete formation, they can be beaten.

31 If the frontline soldiers and backup weaponry are not evenly arrayed in the forefront of the battle formation, they can be beaten.

32 If they worry so much about the front in battle that they leave the rear open, or they worry so much about the rear that they leave the front open, or worry so much about the left that they leave the right open, or worry so much about the right that they leave the left open—if they have any worry in combat, they can be defeated.

If a city is in a marshy area without high mountains or deep canyons, and yet it abuts upon hills on all four sides, it is a strong city, not to be besieged. If their army is drinking running water—that is, water from a live source—it is not to be besieged. If there is a deep valley in front of the city and high mountains behind it, it is a strong city, not to be besieged. If the city is high in the center and low on the outskirts, it is a strong city, not to be besieged. If there are joining hills within the city precincts, it is a strong city, not to be besieged.

If an encamped army rushes to its shelters, there is no large river encircling them, their energy is broken down and their spirits weakened, then they can be attacked. If a city has a deep valley behind it and no high mountains to the left and right, it is a vulnerable city and can be attacked. If the surrounding land is arid, a city is on barren ground and can be attacked. If the troops are drinking brackish water, or stagnant or stale water, they can be attacked. If a city is in a large swampy area with no large valleys, canyons, or abutting hills, it is a weak city and can be attacked. If a city is between high mountains and has no large valleys, canyons, or adjoining hills, it is a weak city and can be attacked. If a city has a high mountain in front of it and a large valley behind it, so that it is high in front and low in back, it is a weak city and can be attacked.

. . . ! When reinforcements arrive, they can be beaten too. So a general rule for military operations is that groups over fifteen miles apart cannot come to each others' rescue—how much less when they are at least thirty and up to a hundred or more miles apart! These are the extreme limits for grouping and spacing battalions.

Therefore military science says that if your supplies do not match those of opponents, do not engage them for long; if your numbers do not match up to those of opponents, do not get embroiled with them; . . . if your training does not match that of opponents, do not try to contest them where they are strongest. Once these five assessments are clear, a military force may act freely.

So a military operation . . . heads for opponents' strategic factors. First, take their fodder. Second, take their water. Third, take the fords. Fourth, take the roads. Fifth, take the rugged ground; sixth, take the level ground. . . . Ninth, take what they consider most valuable. These nine seizures are ways of taking opponents.

... The concentrated prevail over the scattered, the full prevail over the empty, the swift prevail over the slow, the many prevail over the few, the rested prevail over the weary.

Concentrate when there is reason to concentrate, spread out when there is reason to spread out; fill up when there is reason for fullness, empty out when there is reason for emptiness. Go on the byways when there is reason to go on the byways, go by the highways when there is reason to go by the highways; speed up when there is reason to speed, slow down when there is reason to slow down. Mass in large contingents when there is reason for huge masses, group in small contingents when there is reason for small groups. Relax when there is reason to relax, work hard when there is reason to work hard.

Concentration and scattering interchange, fullness and emptiness interchange, byways and highways interchange, swiftness and slowness interchange, many and few interchange, relaxation and labor interchange. Do not confront the concentrated with concentration, do not confront the scattered by scattering. Do not confront the full with fullness, do not confront the empty with emptiness. Do not confront speed with speed, do not confront slowness with slowness. Do not confront many with many, do not confront few with few. Do not confront the relaxed when you are in a state of relaxation, do not confront the weary when you are in a condition of weariness.

The concentrated and the spread-out can oppose each other, the full and the empty can oppose each other, those on byways and those on highways can oppose each other, the fast and the slow can oppose each other, the many and the few can oppose each other, the rested and the tired can oppose each other. When opponents are concentrated, you should therefore spread out; when they are full, you should therefore be empty. When they go over the byways, you should therefore take the highways. When they speed, you should therefore go slowly. When they are many, you should therefore use small contingents. When they are relaxed, you should therefore labor.

The pattern of heaven and earth is to revert when a climax is reached, to wane on waxing full. [The sun and moon, yin and yang,] are examples of this. There is alternate flourishing and dying out; the four seasons exemplify this. There is victory, and there is failure to prevail; the five elements exemplify this. There is birth, and there is death; all beings exemplify this. There is capacity and there is incapacity; all living creatures exemplify this. There is surplus, and there is deficiency; formation and momentum exemplify this.

So whoever has form can be defined, and whoever can be defined can be overcome. Therefore sages use what is overwhelming in all things to overcome all things; therefore their victories are unstoppable and inexhaustible. Warfare is a matter of formal contest for victory. No form is impossible to overcome, but no one knows the form by which victory is obtained.

▪ ▪ To use what is overwhelming in all things to overcome all things means to use natural forces and intrinsic momenta to accomplish a task. This is why intelligent leadership is essential to certain victory no matter how much raw force is available.

No form is impossible to overcome because formations have inherent laws or patterns whose developments and movements can be predicted. No one knows the form by which victory is obtained in a general sense because there is no set form that will guarantee victory, and in a particularized sense because the specific form that wins in a given case prevails because of its inscrutability to the opposition.

Changes of form and victory are infinite, coterminous with heaven and earth; they could never be fully written down. Form is a matter of using whatever is superior to win. It is impossible to use the best of one form to overcome all forms. Therefore control of forms is one, but the means of victory cannot be only one.

■ ■ Adaptability is a perennial keynote to strategic thinking, but if the central integrity of leadership is compromised, whether in capacity or purpose, adaptability can degenerate into pliability, fragmentation, or dissipation of energy. By the same token, if the central integrity of leadership is lacking in intrinsic strength and the leadership resorts to dogmatic ideological and authoritarian ways to compensate, flexibility of thought and action are sacrificed to stabilization of a fixed structure, finally leading to inability to safeguard the foundation of the structure no matter how stable its internal dimensions may remain.

Therefore experts at warfare see the strengths of opponents, and thereby know their weaknesses; seeing their deficiencies, they thereby know their surpluses. They see victory as clearly as they see the sun and moon; their attainment of victory is like water overcoming fire.

To respond to a form with a form is directness, to respond to form without form is surprise. Directness and surprise are endless, having distinct places. Organize your

divisions by surprise strategy, control others by the five elements, battle them with . . . When the divisions are determined, then there is form; when a form is determined, then there is definition. . . .

Sameness is inadequate to attain victory; therefore difference is used, for surprise. Therefore stillness is surprise to the mobile, relaxation is surprise to the weary, fullness is surprise to the hungry, orderliness is surprise to the unruly, many are a surprise to the few. When the initiative is direct, holding back is surprise; when a surprise attack is launched without retaliation, that is a victory. Those who have an abundance of surprises excel in gaining victories.

■ ■ Surprise tactics are valued to thwart accurate anticipation of your movements on the part of opponents. Continuous fluidity is needed, because "Surprise becomes conventional, convention becomes surprise."

So when one joint aches, all hundred joints are disabled, because they are the same body. When the vanguard is defeated and the rear guard is ineffective, it is because they are in the same formation. Therefore in battle configurations, . . . the rear should not overtake the front, and the front should not trample the rear. Forward movements should follow an orderly course, and withdrawal should return in an orderly manner.

■ ■ Forces and resources are organized for maximal efficacy on the offensive and minimal vulnerability on the defensive. The art of organization, from this point of view, is to enable different units of power to operate in concert and also independently, without these two capacities interfering with each other in practice.

When people obey rules without rewards or punishments, these are orders that the people can carry out. When the high are rewarded and the low punished, and yet the people do not obey orders, these are orders that the people are unable to carry out. To get people to fly in the face of death without turning on their heels, in spite of poor management, is something even a legendary hero would find hard to do; so to put this responsibility on ordinary people is like trying to get a river to flow in reverse.

Therefore in battle formations, winners should be strengthened, losers should be replaced, the weary should be rested, the hungry should be fed. Then the people will only see the enemy, and won't see death; they will not turn on their heels even though they tread on naked blades. So when flowing water finds a course, it can even wash away boulders and snap boats in two; when people are employed in a manner consistent with their nature, then orders are carried out like a flowing current.

LEADERSHIP, ORGANIZATION, AND STRATEGY:

▪ How Sun Tzu and Sun Tzu II Complement Each Other ▪

Sun Bin the Mutilated was a lineal descendant of the famous **Sun Wu the Martialist,** whose *Art of War* is perhaps the best known of the classics of strategy. In 1972 a hitherto unknown version of **Sun Wu**'s work was discovered in an archaeological find at Silver Sparrow Mountain in China's Shandong Province. This version of **Sun Wu**'s *Art of War* predates the traditional commentaries through which this classical text is ordinarily studied. Although **Sun Wu** and, to a lesser extent, **Sun Bin,** have long been known to history, recent developments have made them both new discoveries.

An academic attempt to translate the newly discovered *Art of War* has been made, unfortunately without success, being based on the erroneous belief that the Chinese worldview lacks intelligibility and predictability. Since all strategy (and language, for that matter) depends on intelligibility and predictability, a representation of strategic literature as lacking these factors is, quite naturally, inherently flawed and intrinsically misleading.

In any case, having found that academic work of no value in this connection, for purposes of comparison with **Sun Bin** I draw on my own original unpublished translation of the newly rediscovered text of **Sun Wu**'s *Art of War.*

Three essential features of tactical formulations stand out in both **Sun Wu** and **Sun Bin: leadership, organization,** and **strategy. Leadership** is necessary to the cohesion and direction of organization, and to the election and implementation of strategy. **Organization** is needed to be effective on a large scale; and **strategy** is needed to plan the functional economy of action undertaken by the organization.

The similarities and differences between the tactical science of **Sun Bin** and that of his predecessor **Sun Wu** are clearly apparent, and follow predictable patterns. **Sun Wu,** the elder tactician, tends to be more **summary** and more **abstract;** **Sun Bin,** the successor, is inclined more toward **detail** and **concreteness.** When viewed together in their essences, therefore, the complementary designs of these two major strategists yield a fuller picture of the foundations of tactical thinking.

Leadership is without question the major issue underlying all strategic science, inasmuch as it represents direction and purpose in both ideological and practical domains. **Sun Wu** defines the basic pillars of good leadership in terms of **five requirements: knowledge, trustworthiness, humaneness, valor,** and **strictness.** This is a more concentrated version of the formulation given in the earlier classic, *Six Strategies,* which refers to the qualities of **humaneness, justice, loyalty, trustworthiness, courage,** and **strategy** as the "six defenses" that a leader, or an elite corps, should command and embody to safeguard agriculture, industry, and trade.

Sun Bin also emphasizes requirements in leadership similar to those enumerated by his predecessors:

One who leads a militia with inadequate intelligence is conceited. One who leads a militia with inadequate courage has an inflated ego. One who leads a militia without knowing the Way and does battle repeatedly without being satisfied is surviving on luck.

These parameters might be summarized as knowledge, valor, wisdom, and modesty.

Sun Bin notes the dominant flaws of character in those who are lacking the essential qualities of leadership to underscore the pragmatic nature of these requirements. He also goes further into specifics to illustrate those factors of leadership that lead to success and those that lead to failure. He says,

> There are five conditions that always lead to victory. Those who have authorized command over a unified power structure are victorious. Those who know the Way are victorious. Those who win many cohorts are victorious. Those whose close associates are in harmony are victorious. Those who take the measure of enemies and size up difficulties are victorious.

Qualities conspicuously absent in classical descriptions of good warriors and good leaders are bloodthirstiness, violence of temper, and overweening ambition. **Sun Bin** also said:

> Those who enjoy militarism, however, will perish; and those who are ambitious for victory will be disgraced. War is not something to enjoy, victory is not to be an object of ambition.

The primary practical reason for this warning is explained by the elder master, **Sun Wu:**

> Those not completely aware of the drawbacks of military action cannot be completely aware of the advantages of military action.

Therefore **Sun Bin** outlines sources of defeat in strategic operations:

> There are **five things that always lead to failure**. Inhibiting the commander leads to failure. Not knowing the Way leads to failure. Disobedience to the commander leads to failure. Not using secret agents leads to failure. Not winning many cohorts leads to failure.

Pursuing a similar analysis of failures of leadership, the elder master, **Sun Wu**, in accordance with his dictum that "The considerations of the wise include both profit and harm," also outlines what he calls **five dangers in military leaders**, which may be summarized as follows:

> Those who will fight to the death can be killed.
> Those intent on survival can be captured.
> Those quick to anger are vulnerable to contempt.
> Purists are vulnerable to shame.
> Emotional humanitarians are vulnerable to anxiety.

Typically more detailed than his predecessor, **Sun Wu**, **Sun Bin** devotes several chapters to outlining the qualifications and requirements of leadership. The various attributes of leadership are pictured as parts of the body, all of them forming an integral whole: **Justice**, from which derives **dignity** and thence **charisma**, is the **head of warriorship**. **Humaneness**, which encourages **effectiveness**, is the **gut of warriorship**. **Integrity**, as a foundation for **power**, brings out the best in armies, so it is the **hand of warriorship**. Trustworthiness, which fosters **obedience**, from which derives **unity**, is the **foot of warriorship. Intelligence** fosters **resolution**, which is the **tail of warriorship.**

Sun Bin also outlines failings and losses in commanders at considerable length. Failings in commanders include

> considering themselves capable of what they are unable to do, arrogance, ambition, greed, impulsiveness, slowness, cowardice, weakness, unreliability, irresoluteness, laxity, laziness, viciousness, egocentricity, unruliness.

While the function of the leader—that is, to impart order to the action of a group—naturally requires certain capacities in the person of the leader, the effective power of the leader to direct an organization also depends on the structural integrity or order of the organization. **Sun Wu** summarizes order in these terms: **"Order involves organizational structure, chain of command, and logistics."** The importance of preparing functional bases of operation is also stressed by the later master, **Sun Bin: Intelligent governments and commanders with knowledge of military science must prepare first; then they can achieve success before fighting.**

To achieve success before fighting is to outdo competitors in strategic advantages, including qualities of leadership and personnel, and integrity of organizational structure.

The importance of integrity in the order is made abundantly clear in the classic of **Sun Wu,** where he says,

> **When order is consistently practiced to educate the people, then the people are obedient. When order is**

not practiced consistently to educate the people, then the people are disobedient. When order is consistently practiced, that means it is effective for the group.

The principle that the operative order must be consistent with the effective character and capacity of the group is also emphasized by **Sun Bin,** who says,

When people obey rules without rewards or punishments, these are orders that the people can carry out. When the high are rewarded and the low punished, and yet the people do not obey orders, these are orders that the people are unable to carry out.

The strategic importance of order is the intensive exertion of force or capacity that order makes possible. **Sun Wu** explains it in this way: "**What normally makes managing a large group similar to managing a small group is a system of order.**" The facilitation of intensive exertion is forcefully illustrated by **Sun Bin** in these terms: "**When all is in order, the battle line moves with a shocking commotion.**" To represent the inner cohesion of the organization, by which the integrity of the order is maintained, **Sun Bin** again uses the image of a body, implying that the total integrity and discipline of the whole order depends on the personal integrity and discipline of each individual in the organization: "**When one joint aches, all hundred joints are disabled, because they are the same body.**"

The actualization of an effectively unified order is thus naturally a matter of critical concern. Primary emphasis is placed on the moral and intellectual character of leadership because this unification cannot be attained by simple fiat. According to **Sun Wu,** the attainment of objective organizational integrity depends on the realization of subjective organizational unity:

> If soldiers are punished before an emotional bond has formed [with the leadership], they will not be obedient, and if they are not obedient they are hard to direct. If penalties are not enforced once this emotional bond has formed, then the soldiers cannot be directed. So unite them culturally and unify them militarily; this is considered the way to certain victory.

This was the concern of the **king** who asked, **"How can I get my people to follow orders as an ordinary matter of course?"** To this **Sun Bin** replied, **"Be trustworthy as an ordinary matter of course."** The need for correspondingly effective objective order does, nevertheless, remain imperative; Sun Wu says: **"Whether there is order or unruliness depends on the operative logic of the order."**

In terms of organizational structure, the logic of an operation depends on the recognition and employment of individual capacities in such a way as to maximize their efficiency within the body of the whole, as noted by **Sun Bin** when he says, **"When setting up official posts, you should do so in a manner appropriate to the individual."**

The purpose of this selectivity, of course, is not to fulfill the ambitions of individuals irrespective of the welfare of the group, but to enhance the internal harmony and therefore survival value of the organization, as **Sun Bin** explains: On the one hand, **"If their orders are not carried out and**

their troops are not unified, commanders can be beaten";
while on the other hand, Sun Bin also adds,

> When flowing water finds a course, it can even wash
> away boulders and snap boats in two; when people
> are employed in a manner consistent with their na-
> ture, then orders are carried out like a flowing cur-
> rent.

The qualifications of leadership and the requirements of
order apply, moreover, to every step on the chain of command, from the top commander in charge of the whole
group to the individual in charge of personal performance.
Sun Wu said,

> Those who press forward without ambition for fame
> and retreat without trying to avoid blame, who only
> care for the security of the people and thus are in
> harmony with the interests of the social order, they
> are treasures of the nation.

Subordination of selfish ambition or personal vanity to the
welfare of the group does not deny but rather affirms the
worth of the individual, because the proper combination of
teamwork and individual responsibility is what gets the job
done. The same basic principles are also echoed by Sun Bin:

> Acting with integrity is a rich resource for warriors.
> Trust is a distinguished reward for warriors. Those
> who despise violence are warriors fit to work for
> kings.

In addition to character and organizational ability, the capacity for intelligent planning is essential to leadership. In the words of Sun Wu: **"Those who do not know the plans of competitors cannot enter capably into preliminary negotiations."** Skill in tactical thinking is considered normal for leaders—not a product of cunning artifice, but a natural application of intelligence to the realities of life as it is. **Sun Bin** explains it this way:

> **Fangs and horns, claws and spurs, harmonizing when pleased, fighting when angry—these are in the course of nature, and cannot be stopped. Therefore those who have no natural defenses plan strategically for themselves; this is the business of wise leaders.**

The essence of strategic thinking, the pivot on which tactical action revolves, is situational adaptation, as indicated by the elder master, Sun Wu: **"Leaders who have mastered the advantages of comprehensive adaptation to changes are those who know how to command militias."**

It is this ability to adapt to changes, furthermore, that allows the warrior to remain unruffled in the midst of chaotic upheaval, as **Sun Wu** observes: **"Masters of military affairs move without confusion, mobilize without exhaustion."** In this way the actions and measures taken by the leader can be based on objective response to the situation, unaffected by subjective emotions, unfazed by the pressures of the moment, as **Sun Bin** says: **"Let nothing seduce you, let nothing anger you."** The elder master, **Sun Wu**, remarks: **"To face confusion with composure and face clamor with calm is mastery of heart."**

By remaining calm yet alert, uncaptivated and unper-turbed, the leader can concentrate mental energies on es-sential tasks, and not be sidetracked by ambient unrest. Conversely, skillful focus also enables the leader to be that much less distracted and so much the more serene and un-ruffled. Sun Bin says, **"When articulating direction and establishing a standard of measure, only focus on what is appropriate."**

This helps to alleviate internal unrest in the ranks by re-lieving the minds of subordinates from unnecessary con-cerns; and it also serves to help maintain security by having people usefully occupied while keeping future plans in re-serve, to be revealed only at the appropriate time. As **Sun Wu** describes it, **"The affairs of military commanders are kept inscrutable by quiet calm."**

The inner inaccessibility of the leader is, of course, strate-gic, and must not translate into aloofness and unconcern, for then it would endanger rather than safeguard the security of an organization or an operation. As **Sun Bin** says, **"When you know soldiers are trustworthy, don't let others alien-ate them."** Even while maintaining a hidden secrecy, the leadership must be intimately acquainted with conditions within and without the organization, as illustrated in the often quoted dictum of **Sun Wu**, so simple yet so telling: **"Knowing others and knowing yourself, victory will not be imperiled."**

The meaning of "knowing," so complex and so critical in the context of strategic thinking, may refer or allude in a given case to any one or more of numerous diverse yet more or less indirectly related phenomena, including information, misinformation, disinformation, and censorship—and, in addition, understanding, misunderstanding, illusion, and deception.

The whole science of deliberate construction of these forms and shadows of "knowledge," and the manipulation of their specific interrelationships, is crucial to tactical action at its most sophisticated level. This is the underlying fact that has given rise to the famous dictum of **Sun Wu**, which is itself so often misrepresented, misconstrued, and misunderstood, that **"Warfare is a path of subterfuge."**

Whenever either of the masters Sun involves himself in a discussion of concrete tactics in actual situations, it becomes clear that the notion of warfare as a path of subterfuge, the practice of seizing control of the opponent's very thoughts and perceptions, underlies the whole science of situational mastery and effective surprise.

Sun Wu says, **"Make a show of incompetence when you are actually competent, make a show ineffectiveness when you are in fact effective."**

Sun Bin elucidates,

Let them think you lack resolve, feign lack of ability, and appear to have a defeatist attitude, so as to seduce them into arrogance and laziness, making sure they do not recognize the real facts. Then, on this basis, strike where they are unprepared, attack where they are not defending, pressure those who have slacked off, and attack those who are uncertain or confused.

One reason for the use of surprise tactics is that they are ordinarily more economical than conventional tactics, insofar as they are designed to strike at points of least resistance. To obtain the greatest advantage with the least embroilment is one of the key arts of war.

Defensive maneuvering is thus more than defensive. Not only is it a means of storing energy, it is a way to spy out the intentions and abilities of opponents. **Sun Wu** advises, **"Even when you are solid, still be on the defensive; even when you are strong, be evasive."**

Following up on this idea, **Sun Bin** adds: **"Do not pursue opponents in flight,"** for that would expend precious energy and also expose one's own position and capacity.

These conservative and even defensive maneuvers are bases for attack, which like defense begins with the mental aspect of warfare. **Sun Wu** proposes tactics that radically minimize one's own expenditures while putting the opponent at maximum disadvantage: **"Use anger to make them upset, use humility to make them arrogant."** Following up on this, **Sun Bin** gives some further advice on a convenient way to achieve the latter effect: **"Disarray troops in confused ranks, so as to make the other side complacent."**

The underlying idea is to put off on the enemy as much of the burden of warfare as possible, while reserving oneself intact. This is why the moral philosophy of warfare from which these texts arise is fundamentally nonaggressive and is ethically based on response rather than initiative. From moral philosophy, this is translated directly into practical strategy, as illustrated by **Sun Wu** in his tactical dictum, **"Tire them while taking it easy, cause division among them while acting friendly,"** and echoed by **Sun Bin's** advice: **"You eat to your fill and wait for the enemy to starve; you stay put comfortably and wait for the enemy to tire; you keep perfectly still and wait for the enemy to stir."**

The power accumulated by the practice of maximum economy achieved by secrecy and reserve is enhanced by the ability to compromise the power of opponents. Since a direct approach to diminishing the enemy's force would be most costly, again the scientific approach is preferred, as **Sun Bin** illustrates in his pivotal dicta: **"Confuse them and split them up,"** and **"Those skilled at warfare are skilled at trimming enemies down and cutting their forces apart."** Strategic preference for this approach is also evident in the advice of **Sun Wu,** when he says, **"A superior military operation attacks planning, the next best attacks alliances."** These are primary ways of splitting up opponents' forces. **Sun Bin** adds, **"Attack them as they set up a front, split them up as they organize,"** explaining the logic of such maneuvers in these terms:

> **Those who are able to split up others' armies and control others' forces are adequate even with the smallest quantity; those who are unable to split up others' armies and control others' forces are inadequate even if they have several times the firepower.**

One aspect of force splitting is the deliberate dividing of attention. This type of strategy is common and general in application, as in the advice of **Sun Wu** to **"Strike where they are unprepared, emerge when they are least anticipating it,"** and the counsel of **Sun Bin** to **"Attack where they are unprepared, act when they least expect it."**

This sense of opportune time and place is, furthermore, not simply a matter of the enemy's concrete preparedness, but also a question of mental energy and morale. **Sun Wu** advises, "Good warriors avoid keen spirits, instead striking enemies when their spirits are fading and waning." To this may be added, pursuant to the already stipulated need to know both others and self in order to act effectively, Sun Bin's essential caveat: "Act only when prepared."

The critical discernment of power configurations, of the relationships between one's own states and those of opponents, is underscored in Sun Wu's summary of this aspect of tactical strategy: "The ancients who were skilled in combat first became invincible, and in that condition awaited vulnerability on the part of enemies." The economy of this approach is strongly emphasized in Sun Bin's suggestion that the best preparation is that which enables you to avoid embroilment in persistent hostilities: "Intelligent governments and commanders with knowledge of military science must prepare first; then they can achieve success before fighting."

Preparation sufficient to secure victory in advance requires knowledge of environmental factors. Both the elder master, Sun Wu, and his successor, Sun Bin, emphasize the importance of prior knowledge. Sun Bin says, "It is imperative to know what ground is viable and what ground is deadly; occupy the viable and attack the deadly." Here Sun Wu goes into some detail, specifying the appropriate measures to take on particular grounds:

On a ground of disintegration, do not fight. On shallow ground, do not halt. On a ground of contention, do not attack. On a ground of intercourse, do not get cut off. On axial ground, make alliances. On deep ground, plunder. On bad ground, keep going. On surrounded ground, plan ahead. On deadly ground, fight.

The timing of an operation is as critical as the field of operation. Sun Bin says, "Fight only when you are sure to win, without letting anyone know." Concentration and targeted release of power require inward certainty and outward security for maximum efficiency, as illustrated by Sun Wu's dictum that "Crushing force is due to timing and control."

It is not enough to have the power; it must be focused and directed, as Sun Bin says: "When your forces are larger and more powerful, and yet you still ask about how to employ them, this is the way to guarantee your nation's security."

Reliance on superior force is not merely risky but intrinsically costly, since it depends on the logic of expenditure. The need to devise effective structures through which force can be concentrated and given aim is an established strategic priority, but it is complicated by the fact that even an effective formulation loses its edge once it has become routine. As Sun Wu points out, "Usually, battle is engaged in a conventional manner but is won by surprise tactics. . . . Surprise and convention give rise to each other in cycles."

Once a tactic has become habitual, its effectiveness is lost; the enemy can see through the strategy and be prepared with a counter maneuver. Thus Sun Bin warns, "Whoever has form can be defined, and whoever can be defined can be overcome." This is not only a basic principle of defensive warfare; it is also fundamental to offense, as explained by Sun

Wu in these terms: "If you induce others to adopt a form while you remain formless, then you will be concentrated while the enemy will be divided."

Formlessness, which also means fluidity of form, is thus not merely defensive but is also effective as an offensive posture. Sun Bin says, "A militia is not to rely on a fixed formation," because fixation leads to exhaustion, paralysis, and loss of opportunity. Thus Sun Wu teaches,

> The consummate formation of a militia is to reach formlessness. Where there is no specific form, even deeply placed agents cannot spy it out; even the canny strategist cannot scheme against it.

This does not mean that there is no form whatsoever, but that there is no fixed form, as Sun Bin explains:

> Form is a matter of using whatever is superior to win. It is impossible to use the best of one form to overcome all forms. Therefore control of forms is one, but the means of victory cannot be only one.

The idea that fluid adaptability underlies successful strategy is made quite clear in the corresponding dictum of Sun Wu: that "A militia has no permanently fixed configuration, no constant form. Those who are able to seize victory by adapting to opponents are called experts."

The need for flexibility is emphasized in strategic literature partly because variation is in the nature of things, as Sun Wu notes in remarking that "No element is always dominant, no season is always present." The ability to change tactics is not only necessary for adaptation to external changes in circumstances; it is strategically necessary in order to baffle opponents. **Sun Wu** brings this out when he says, **"The task of a military action is to unobtrusively deceive the minds of enemies,"** and **Sun Bin** confirms that **"Experts drive their enemies to their wits' ends."**

Thus **Sun Wu** advises,

> **When the enemy presents an opening, be sure to penetrate at once. Preempt what the enemy prefers, secretly anticipating him. Act with discipline and adapt to the opposition in order to settle the contest.**

This fluid skillfulness is described by **Sun Bin** in these terms: **"To respond to a form with a form is directness, to respond to form without form is surprise."**

For opponents of the expert, the question of whether one uses form or formlessness, convention or surprise, ultimately becomes a formless surprise in itself. For the tactician, it is simply a matter of what will work effectively; as **Sun Wu** says: **"Do not mobilize when it is not advantageous, do not act when it is not productive, and do not fight when not imperiled."**

After all factors have been considered, and the logic of the operation is clear, the decision to mobilize can be approached with intelligence. Then, if strategic necessity calls for it, subterfuge is the essence of the art of war. As **Sun Wu** says, in his colorful description of tactical surprise, **"At first you are like a virgin girl, to whom the enemy opens his door. Then you are like a jackrabbit on the loose, which the enemy cannot keep out."**

The prominence of subterfuge and deception in the techniques of classical strategists such as **Sun Wu** and **Sun Bin** often gives the impression of thoroughgoing ruthlessness. What must be remembered is that tactical action, as understood by these ancient thinkers, is not only considered from a material point of view, but also from psychological and philosophical points of view. Thus strategy is legitimately conceived in the aftermath, and in the reflection, of moral and ethical consideration.

This critical factor is immediately evident in **Sun Wu's** own introduction to the subject: **"War is a national crisis; it is necessary to examine the grounds of death and life, and the ways to survival and extinction."** The premise that warfare is justified by its moral ground, not by its outcome, is often overlooked by those who focus only on strategy per se without keeping the ethical dimension of the context ever present in the background.

This moral and ethical basis underlying the tradition of **Sun Wu** and **Sun Bin** can be found fully expressed in the classic *Six Strategies,* which is a basic source book for all the great works of this type. This classic is attributed to a sage of the twelfth century B.C.E., a teacher of kings from whose vast body of work derive the main sources of Chinese culture and civilization.

It is in the voluminous *Six Strategies* that both ethical and pragmatic aspects of statecraft and strategy can be clearly seen, foreshadowing the later teachings of **Sun Wu** and **Sun Bin** on strategic factors of leadership, order, and command.

The original teachings of the *Six Strategies* on the subject of leadership include the whole person, from character, mentality, attitude, and conduct as an individual, to manners and

techniques proper to professional management and leadership skills:

> Be calm and serene, gentle and moderate. Be gen-
> erous, not contentious; be openhearted and even-
> minded. Treat people correctly.
>
> Don't give arbitrary approval, yet don't refuse out
> of mere contrariness. Arbitrary approval means loss of
> discipline, while refusal means shutting off.
>
> Look with the eyes of the whole world, and there
> is nothing you will not see. Listen with the ears of the
> whole world, and there is nothing you will not hear.
> Think with the minds of the whole land, and there
> will be nothing you do not know.
>
> If you are lazy even when you see there is good to
> be done, when you are hesitant even though the time
> is right, if you persist in something knowing it is
> wrong, this is where the Way halts. When you are
> flexible and calm, respectful and serious, strong yet
> yielding, tolerant yet firm, this is where the Way
> arises.
>
> When duty prevails over desire, this results in
> flourishing; when desire prevails over duty, this results
> in perishing. When seriousness prevails over laziness,
> this results in good fortune. When laziness prevails
> over seriousness, this results in destruction.

Earlier mention was made of the so-called six defenses listed
in the *Six Strategies,* which are analogous to **Sun Wu's** parameters for leadership. These "six defenses" are actually qualities and capabilities of capable commanders: humaneness,

justice, loyalty, trustworthiness, courage, and strategy. The *Six Strategies* also lists ways of choosing people for these six defenses:

Enrich them and see if they refrain from misconduct, in order to prove their humaneness.

Ennoble them and see if they refrain from hauteur, in order to prove their sense of justice.

Give them responsibilities and see if they refrain from autocratic behavior, in order to prove their loyalty.

Employ them and see if they refrain from deceit, in order to prove their trustworthiness.

Endanger them and see if they are unafraid, in order to prove their courage.

Burden them and see if they are unflagging, in order to prove their strategic approach to problems.

The principles of order elucidated in *The Six Strategies* are, like the principles of leadership, forerunners of the concepts of organization utilized by both **Sun Wu** and **Sun Bin.**

Some of the most powerful of these are the diagnostic principles by which defects in a system can be identified. Among these are the so-called **Six Robbers** and **Seven Destroyers.**

The **Six Robbers** are:

Officials who build huge mansions and estates and pass their time in entertainment, to the detriment of the integrity of leadership.

Workers who don't work, but go around getting into others' business, disrupting order.

Officials who have cliques that obscure the good and wise and thwart the enlightened.

Ambitious officers independently communicate with leaders of other outfits, without deference to their own leaders.

Executives who disregard rank and look down on teamwork, and are unwilling to go to trouble for employers.

Strong factions who overpower those who are weak and lacking in resources.

The **Seven Destroyers** are:

Those who lack intelligent tactical strategy but are pugnacious and combative out of ambition for rewards and titles.

Self-contradicting opportunists, pretenders who obscure the good and elevate the bad.

Those who put on the appearance of austerity and desirelessness in order to get something.

Those who pretend to be eccentric intellectuals, putting on airs and looking on the world with aloof contempt.

The dishonest and unscrupulous who seek office and entitlement by flattery and unfair means, who display bravery out of greed for emolument, who act opportunistically without consideration of the big picture, who persuade leaders with tall tales and empty talk.

Those who comprise primary production by needless luxury.

Those who use supposed occult arts and superstitious practices to bewilder decent people.

Because selection and employment are considered part of the overall task of management, the question arises as to why there may be no effective results even if the leadership tries to promote the worthy. The answer provided by the *Six Strategies* is that in such cases, promotion of the worthy is more form than reality, going on the basis of vulgar popularity or social recommendation and not finding really worthy people:

> If the leadership considers the popular to be worthy and the unpopular to be unworthy, then those with many partisans get ahead, while those with few partisans fall behind. If so, then crooks will be everywhere, obscuring the worthy; loyal administrators will be terminated for no wrongdoing, while treacherous bureaucrats will assume rank by means of false representation.